DATE DUE

MAY 0 1 2006		
JUN 0 4 2006		Anual Date

DEMCO 38-296

In Search of Sustainable Water Management

In Search of Sustainable Water Management

International Lessons for the American West and Beyond

Edited by

Douglas S. Kenney

Natural Resources Law Center
University of Colorado School of Law, USA

Edward Elgar
Cheltenham, UK • Northampton, MA, USA

Published by
Edward Elgar Publishing Limited
Glensanda House
Montpellier Parade
Cheltenham
Glos GL50 1UA
UK

Edward Elgar Publishing, Inc.
136 West Street
Suite 202
Northampton
Massachusetts 01060
USA

A catalogue record for this book
is available from the British Library

Library of Congress Cataloguing in Publication Data
In search of sustainable water management : international lessons for the American
   West and beyond / edited by Douglas S. Kenney.
       p. cm.
     Includes bibliographical references.
       1. Water-supply–West (U.S.)–Management. 2. Water resources development–
   West (U.S.) 3. Sustainable development–West (U.S.) 4. Water-supply–
   Management–Case studies. 5. Water resources development–Case studies. 6.
   Sustainable development–Case studies. I. Kenney, Douglas S., 1964-

   HD1695.W32I62 2005
   333.91'16'0978–dc22

                                                                      2004058625

ISBN 1 84376 944 1 (cased)

Typeset by Manton Typesetters, Louth, Lincolnshire, UK.
Printed and bound in Great Britain by MPG Books Ltd, Bodmin, Cornwall.

Contents

List of tables   vi
Contributors   vii
Preface   xiii

1   Water policy and cultural exchange: transferring lessons from
    around the world to the western United States   1
    James L. Wescoat Jr

2   Roles for the public and private sectors in water allocation:
    lessons from around the world   25
    Charles W. Howe and Helen Ingram

3   Integrating environmental and other public values in water
    allocation and management decisions   69
    David H. Getches and Sarah B. Van de Wetering

4   Protecting indigenous rights and interests in water   102
    David H. Getches and Sarah B. Van de Wetering

5   Transboundary water conflicts and cooperation   131
    Aaron T. Wolf

6   Sustainability and the future of western water law   155
    Lakshman Guruswamy and A. Dan Tarlock

Index   181

List of tables

1 Drawing lessons of difference and similarity   10
2 Diffusion of innovations models   12

Contributors

Don J. Blackmore was Chief Executive of the Murray–Darling Basin Commission, Australia, 1990–94. Prior to that he was Deputy Chief Executive for six years and worked for 15 years as a Civil Engineer with the Rural Water Commission in Victoria. He was also Deputy Chair of the Land and Water Resources Research and Development Corporation, a position that he held from 1990 to 1999. He was a Commissioner on the World Commission on Dams with a mandate to review the development effectiveness of large dams and criteria for the future investment in dams. He became a Fellow of the Institute of Engineers Australia in 1995 and a Fellow of the Academy of Engineering and Technological Sciences in November 1998. In May 2000 he was awarded the degree of Doctor of Science (honoris causa) by La Trobe University.

Joachim Blatter is an Assistant Professor, Department of Politics and Management, University of Konstanz, Germany. He is the author of a number of journal articles on cross-border regions, and edited (with Helen Ingram) *Reflections on Water: New Approaches to Transboundary Conflicts and Co-operation* (2001).

David Farrier is a Professor of Law and a program manager in the Institute of Conservation Biology and Law at the University of Wollongong, Australia. His published works are in the areas of biodiversity conservation law and policy, water law and policy, integrated natural resource management and the implementation of international conventions (supported by the Australian Research Council). In addition to his several consultancy positions, he is currently a member of the New South Wales Water Advisory Council and the Environmental Consultative Committee, Legal Aid Commission of New South Wales, and is Chairperson of the Shoalhaven/Illawarra Water Management Committee. He has taught law in England, Nigeria and Australia.

David H. Getches is Dean of the University of Colorado School of Law, US, where he also serves as a Professor of Natural Resources Law. In addition to authoring numerous professional articles, books, and casebooks, he is also the founding Executive Director of the Native American Rights Fund (NARF), and from 1983 to 1987 served as Executive Director of the Colorado Department of Natural Resources under Governor Lamm. He has consulted widely concerning water policy and national policies concerning indigenous peoples with governmental agencies and non-governmental organizations throughout the United States and in several foreign countries.

Lakshman Guruswamy is a Professor at the University of Colorado School of Law, US where he specializes in International Environmental Law. His expertise is this field is well established through numerous articles, books, textbooks (e.g., *International Environmental Law in a Nutshell*), and projects, as well as past service as Director of the National Energy–Environment Law and Policy Institute at the University of Tulsa College of Law, his contributions as a Professor at the University of Arizona and University of Iowa Colleges of Law, and his work in the United Kingdom and Sri Lanka. In addition to water-related issues, his experience includes matters of biodiversity protection and environmental dimensions of war and arms control.

Robert K. Hitchcock is a Professor at the University of Nebraska, US. He has conducted fieldwork among the Tyua and Kua Bushman (Basarwa, San) of the east-central Kalahari Desert region of the Republic of Botswana, Somali and Oromo refugees in Somalia, and Swazi women's groups and traditional leaders in Swaziland. He has been involved in the implementation of large-scale rural development projects in Botswana, Somalia, and Swaziland and has done monitoring and evaluation of development programs in Botswana, Lesotho, Malawi, Namibia, Zimbabwe and Zambia. His work seeks to present theoretically robust assessments of development project impacts on the wellbeing of rural poor people, especially hunter-gatherers, pastoralists, small farmers, and rural women. Currently, he is concentrating on applying this expertise to the design and evaluation of natural resource, agricultural and income generating projects. His work concentrates in part upon human rights and community-based resource management strategies.

Charles W. Howe is Professor Emeritus of Economics at the University of Colorado, US. He is a Senior Research Associate in the University's Environment and Behavior Program that he directed from 1986 to 1997. Professor Howe directed the Water Resources Program for Resources for the Future from 1965 to 1970. He has worked on water policy and project design in several countries of East and West Africa, Mexico, Argentina and Indonesia. His major current interests include water marketing, economic impacts of water transfers and urban water pricing.

Helen Ingram is a Professor at the University of California at Irvine, US. Among her nine books and more than eighty articles and book chapters, roughly two-thirds deal with water resources, including a number of studies of transboundary issues. Her coauthored (with Nancy R. Laney and David M. Gillilan) 1995 book *Divided Waters: Bridging the US–Mexico Border* and her 1988 coauthored (with Suzanne L. Fiederlein) article 'Traversing boundaries: a public policy approach to the analysis of foreign policy' exemplify some of the intellectual roots of her chapter in this volume. Her research interests include transboundary national resources, particularly on the US–Mexican border, water resources and equity, public policy design and implementation, and the impact of policy upon democracy and public participation.

Jeffrey Jacobs is a Senior Staff Officer with the Water Science and Technology Board, National Research Council, US. In this position, he directs a variety of research projects, recently focusing on issues including planning by the US Army Corps of Engineers, water management in the Missouri Basin, and water privatization. Much of his international work has focused on water management in the Mekong Basin.

Douglas S. Kenney has been a Senior Research Associate with the Natural Resources Law Center, University of Colorado School of Law, US, since 1996. His research primarily focuses on western water issues, particularly issues of allocation, regional planning and dispute resolution, and law and policy reform. More recently, he has examined the potential impacts of climate change and variability on western water resources as a member of the Western Water Assessment program sponsored by the National Oceanic and Atmospheric Administration. He has served as an advisor and contributor to several other regional, national, and international research projects, including the work of the Western Water Policy Review Advisory Commisision.

Marcus Moench is the President of the Institute for Social and Environmental Transition, Colorado, US. He has extensive experience working with communities, non-government, government and international organizations on water, energy and forest management in South Asia, the Middle East and the western United States. He combines a strong technical background in environmental science, hydrogeology and forestry with training and experience in the design and initiation of management institutions. He recently led the India Water Sector Review, Groundwater Component, and the Yemen Decentralized Management Study for the World Bank.

Mikiyasu Nakayama is the Associate Dean and Professor of the United Graduate School of Agricultural Science, Tokyo University of Agriculture and Technology, Japan. From 1986 to 1989, he served as a program officer in the United Nations Environment Programme (UNEP), where he participated in projects in such international water bodies as the Zambezi and Mekong Rivers and Lake Chad. He has taught water resources management and its international and environmental aspects at the Utsunomiya University, and has served as an advisor and an expert for several United Nations organizations (UNEP, UNCHS, UNCRD, and UNU), as well as for non-governmental organizations such as IUCN and ILEC. His research subjects include the application of satellite remote sensing data for environmental monitoring, using the Geographical Information System (GIS) for environmental management of river and lake basins, employing environmental impact assessment methodologies applicable to involuntary resettlement due to dam construction, and involvement of international organizations in management of international water bodies.

Miguel Ricardo Solanes serves with the United Nations Economic Commission Latin America and the Caribbean (ECLAC) in Santiago, Chile, where he currently is the Interregional Adviser in Water Legislation and Administration. His activities primarily include research and advisory missions, often focused on the use of law to promote environmental protection, resources development, and public utility regulation. He holds a law degree from the University of Mendoza, Argentina, where he has also taught. Additionally, he has conducted research at several US universities – namely, the Kennedy School of Government, Michigan State University, and Colorado State University.

Robyn Tanya Stein is director of Bowman Gilfillan Inc., a Johannesburg, South Africa-based company of attorneys, notaries and conveyancers consisting of 60 partners whose clientele is made up of national and multinational corporations. Among the many professional affiliations she maintains are posts serving as special adviser to Minister Kader Asmal, Minister of Water Affairs and Forestry; as a member of Water Law Review Process Policy and Strategy Team; as a member of the IUCN (World Conservation Union) Commission on Environmental Law; and as a member of several national environmental law associations. She is a part-time lecturer at the University of the Witwatersrand in Environmental Law to LLM and LLB students and a frequent presenter at conferences and seminars.

A. Dan Tarlock is currently Distinguished Professor of Law at the Chicago-Kent College of Law in Chicago, US. Previously, he was a member of the faculty of Indiana University, Bloomington from 1968 to 1982 and has held visiting positions at the universities of Brigham Young, Chicago, Pennsylvania,

Kansas, Michigan, Texas, and Utah. In 1996, he was a Distinguished Foreign Visitor in Residence at the Queensland University of Technology in Brisbane, Australia. He has practiced law in San Francisco, Omaha and Denver, and is an elected member of the American Law Institute. He has consulted widely both in the United States and Australia, Austria, Brazil, Canada, China, Scotland and Germany in the fields of law, domestic and international environmental protection and natural resources management.

Marcos Terena is a native Brazilian tribesman from the Pantanal region, close to the Bolivia and Paraguay border. He founded the first indigenous movement in Brazil, and has represented native interests in a variety of United Nations committees and events, including the 1992 Rio Summit (organizing the World Conference of Native People on Territory, Environment and Development) and the workgroup on indigenous questions in Geneva, Switzerland. He is also a member of the Intertribal Committee (ITC), an indigenous organization with 22 native nations, and was one of the first members of the Amazonic Coalition in Washington. He is a contributor and publishing advisor to the *PNUMA-Terramerica* magazine, and author of a book entitled *The Aviator Indian (O Indio Aviador)*.

Julie Trottier is Thames Water Fellow at St Peter's College Oxford University, UK, where she is based. Her expertise is on the politics of water supply, with special reference to the Middle East and Southern Africa. She was primary contact and the coordinator for the major international workshop in October 2001 at the University, and for the regular academic reading group seminars. Her present research concerns the political/economic/social aspects of water, water development and water management in the Middle East and Southern Africa. Her area of focus in the Middle East has been Israel and the Palestinian Territories, but is also extending now to Turkey and to Saudi Arabia.

Sarah Van de Wetering is a writer/editor and policy consultant on western resource issues, based in Missoula, Montana, US. She helped create and was the editor of *Chronicle of Community*, was a coauthor of *Water in the West* (the final report of the Western Water Policy Review Advisory Commission), and contributed to several books on western water, including: *Searching Out the Headwaters: Change and Rediscovery in Western Water Policy*; *Natural Resources Policy and Law: Trends and Directions*; and *Overtapped Oasis: Revolution or Reform for Western Water*. She was also formerly Associate Director of the Natural Resources Law Center and an attorney for the Sierra Club Legal Defense Fund (San Francisco). She is a graduate of the University of Colorado School of Law.

James L. Wescoat Jr is the head of the Department of Landscape Architecture, University of Illinois at Urbana-Champaign, US. He is currently a member of the National Research Council Water Science and Technology Board. His research focuses on water management and policy issues in the western US and South Asia (primarily Pakistan, India, and Bangladesh). He has conducted research on the geographic logic of Roman, Islamic and US water law; the historical and cultural geography of water development; and the linkages between water policy and landscape planning. His teaching includes courses on international, western, and urban water policy; and water in environmental design at the site scale.

Aaron T. Wolf is an Associate Professor of Geography at Oregon State University, US, whose research and teaching focus on the interaction between water science and water policy, particularly as related to conflict and its resolution. His recent research focuses on issues relating international water resources to political conflict and cooperation, where his training combining environmental science with dispute resolution theory and practice have been particularly appropriate. He has acted as consultant to the US Department of State, the US Agency for International Development, and the World Bank on various aspects of international water resources and dispute resolution. He has been involved in developing strategies for resolving water aspects of the Arab–Israeli conflict, including coauthoring a State Department reference text, and participating in both official and 'track II' meetings between coriparians. He coordinates the Transboundary Freshwater Dispute Database and is an associate editor of the *Journal of the American Water Resources Association*, *Water International*, and *World Water Policy*.

Wang Xi is a Professor at the School of Law at Wuhan University, China, where he is also Deputy Director of the Research Institute of Environmental Law. Additionally, he has also taught abroad in many nations, including the Netherlands, Australia, Canada. Among his many books are *International Environmental Law*, and *Environmental Law of the USA*. A frequent speaker and panelist at national and international conferences, he also directs research projects and consults to UN agencies, national institutions and the European Union (EU). He directed the Project on Law of EU and EU Countries for Greenhouse Gas Reduction, the China-EU Higher Education Exchange Programme (2000) and the Project on Development of International Environmental Law and its Impacts on Modern International Law (1998).

Preface

Water has always been a salient issue in the American West. Mark Twain's familiar summation that "whiskey is for drinkin', water is for fightin'" is an admittedly tired and overused saying, like many revered adages about western water. Yet it nonetheless underscores the notion that our interest in water often runs wider and deeper than the streams themselves – going beyond what even a rigorous classification of economic, environmental, and spiritual values in water can capture to include more basic concerns such as our safety, security, and regional identity, and what, in the final analysis, we are willing to accept as fair and appropriate in our laws, policies, and management regimes. Ultimately, most water issues are not merely about water; that's not why we fight. Rather, most water issues are about values, aspirations, expectations, and, perhaps more importantly, about fears and uncertainties. This common nucleus of concerns unites an otherwise unrelated set of complex and site-specific water issues.

Viewed in this light, water issues in the American West are not so different than those seen elsewhere in the world. To the contrary, many of the most pressing water management problems in the West have obvious international analogs. For example, the depletion of water resources is, sadly, a nearly universal theme. The Colorado and, in this period of extraordinary drought, the Rio Grande cannot maintain a continuous flow of water to the sea, a quality shared by the Nile in Eqypt, the Ganges in Bangladesh, the Yellow in China, and the Amu Dar'ya and Syr Dar'ya in Central Asia. The steady decline of California's Salton Sea is mirrored by the ongoing demise of the Aral Sea, Dead Sea, Sea of Galilee, and Lake Chad, among others. Similarly, groundwater declines in the Central Valley and High Plains (for example, Ogallala region) are not fundamentally different than what is seen in northern China, parts of India, and thousands of other locales.

The water management environments of the West and many distant nations also share many similarities, as evidenced by the following question, asked and answered by Peter Gleick (2002, p. 1) of the Pacific Institute for Studies in Development, Environment, and Security:

> What has a population of nearly 40 million people; an arid and semi-arid climate with mountainous areas where most of the rain falls; serious public opposition to new major dams; the vast majority of water consumed by irrigated agriculture,

much of it on subsidized, low-valued crops; inefficient domestic water use, with a large fraction going to water lawns; conflicts between environmental uses, agricultural uses, and domestic uses; and government planning agencies that haven't yet figured out how they will have to change to meet a changing world? The answer is South Africa. And the Lower Colorado River Basin. And California and Nevada combined. And the Jordan River in the Middle East.

In the American West, as elsewhere, population growth exacerbates longstanding problems of inappropriate water use and management, and underscores the need for improved institutional arrangements. In just the last quarter of the 20th century, the population of the West grew by approximately 32 per cent, compared to a national growth rate of 19 per cent (Case and Alward, 1997). While most of these new residents reside in Southern California, in terms of percentages, it is the states of the Interior West that are undergoing the most significant demographic shifts. Las Vegas, for example, grew by 83 per cent during the 1990s, pulling the State of Nevada to an overall growth rate of 66 per cent in that period (Census Bureau, 2001). These are figures more typically associated with developing nations. This trend is expected to continue; the Census Bureau expects the West to add about one million new residents per year over the next three decades (Census Bureau, 1997).

As Gleick (2002), and others, have noted, one consequence of this population growth is to reduce the amount of water available per capita to satisfy the full spectrum of human needs and desires. Globally in 1850, the freshwater theoretically available per capita was 43 000 cubic-meters/year; today, it is 8000 cubic-meters/year. While this is enough for even the profligate users of the United States (who average about 2000 cubic-meters/year), it is nonetheless a disturbing and inherently unsustainable trend. More people mean a greater need for food production, and thus, a need to continue devoting the lion's share of freshwater supplies globally to irrigated agriculture during a period in which economics and politics demand a reallocation of water to the municipal and environmental sectors.

Equally troubling is the prospect of global climate change, which may shrink the size of the freshwater resource available for human use, in some cases through reduced precipitation or increased evaporation, and perhaps more commonly in the West, through changes in the hydrologic cycle that make our water infrastructure less efficient (Frederick and Gleick, 1999). Under any reasonable future climate scenario, only a small fraction of total water resources will remain as practically available freshwater. As noted by the World Commission on Dams (2000, preface), 'less than 2.5 per cent of water is fresh, less than 33 per cent of fresh water is fluid, less than 1.7 per cent of fluid water runs in streams'. This quantity is only slightly malleable by engineering, technology, and heroic financial investments. Thus, if we are to balance our water budgets, we will need to get better at using what water resources we have already developed, and in findings ways to moderate our growing collective thirst. This is the ubiquitous challenge of sustainable water management – from Angola to Arizona.

These similar challenges to sustainable water management are often over-looked and unappreciated within the water management community, for many reasons. Perhaps most importantly, water systems, and the managers of those systems, operate at local scales, and are largely immune from actions else-where. If events in South Africa, for example, are unlikely to ever create an operational problem for the Metropolitan Water District of Southern Califor-nia, is there truly a compelling need for the huge California utility to look to this distant nation for insights, even if it faces challenges that, conceptually, are similar? This is a legitimate question. Also legitimate is the observation that the water managers of the American West have a different set of opportu-nities and constraints than many of their colleagues in other nations. This is perhaps most evident regarding the public health aspects of water manage-ment. For example, the World Health Organization estimated that for the year 2000 only 60 per cent of the world's population – but 100 per cent of North Americans – had access to water sanitation systems (Gleick et al., 2002, Table 5). Clearly it would be foolish to assume all water managers operate in similar environments.

Perhaps the most important contextual difference between the West and elsewhere is the legal regime for water allocation. Although not completely without precedent or international analog, western water law is clearly unu-sual. At the heart of this law is the doctrine of prior appropriation, based on a simple 'first come/first served' principle easily grasped by anyone who has ever waited in a queue. The system awards the right to use and consume public waters in the West to individual users such as farmers and cities based on a seniority system where the first users of a resource establish perpetual rights to continue that use year after year, with new users entitled only to the water left over after the first user takes their usual entitlement. Those with the oldest uses (*appropriations*) are termed *seniors*, and enjoy the piece of mind of receiving full water supplies in times of shortage while the *juniors*, the late-comers, take their chances. Much like other American institutions, it is a system based on winners and losers, rather than shared sacrifice, driven primarily by individualistic, *laissez-faire* decision processes rather than coor-dinated or centralized planning. As discussed throughout this book, this legal and administrative regime has many important subtleties and variations be-yond this simplistic description; for example, the definition of 'beneficial' (that is, legally recognized) uses is ever-evolving, and the application of prior appropriation has not been extended across state lines, international borders, and to many groundwater reserves. Nonetheless, while these and related details are important and deserving of detailed exploration, it is ultimately the most basic tenets of western water law that shape the political context within which all western water issues are addressed.

This cursory review of similarities and differences leads to a clear conclusion: while the contextual differences between the American West and other nations ensure that it is *not essential* to look abroad for water management insights and lessons, the strong similarities suggest that such an inquiry is likely to be *worthwhile*. This conclusion was given life in June 2002, when the University of Colorado's Natural Resources Law Center – in celebration of twenty years of western water research – focused its annual western water conference at a global scale. *Allocating and Managing Water for a Sustainable Future: Lessons from Around the World* featured approximately 70 international and domestic presenters, highlighting lessons from countries including Angola, Argentina, Australia, Austria, Bangladesh, Bolivia, Botswana, Brazil, Cambodia, Canada, Chile, China, Costa Rica, Cuba, France, Germany, Hungary, India, Israel (and the West Bank), Japan, Laos, Mexico, Namibia, Nepal, Pakistan, Slovakia, South Africa, Switzerland, Thailand, Turkey, United States and Vietnam. This book is a synthesis and outgrowth of that event, featuring chapters authored by key presenters and supplemented with materials from invited panelists, other presenters, and an extremely diverse and well-informed audience.

Chapter 1, authored by Jim Wescoat, heavily draws upon history and academic theory to explore the opportunities, arguments, and mechanisms for transferring lessons between the American West and foreign nations. While major agencies such as the United Nations and World Bank, and publications as diverse as *The Economist* and the *Christian Science Monitor*, race to proclaim water as *the* issue of the twenty-first century, it is important to realize the search for improved water management strategies has a long history. At one time, water managers in the West frequently sought insights from other regions, but the tradition in recent years has been merely to export, not import, lessons. Given the abundance of water problems in the West and the wealth of international innovations, and considered alongside the growing ease of international communication and travel, this American isolationism is troubling, and is certainly contrary to the spirit of this project.

Chapters 2 through 5 focus on four major substantive themes that currently challenge the western water management community, and how these issues are addressed in a variety of international settings. These discussions draw largely from the remarks and written submissions of an esteemed collection of conference panelists, namely Don Blackmore, Joachim Blatter, David Farrier, Bob Hitchcock, Jeff Jacobs, Marcus Moench, Miki Nakayama, Miguel Solanes, Robyn Stein, Marcos Terena, Julie Trottier, and Wang Xi.

In Chapter 2, Charles Howe and Helen Ingram look at the issue of water allocation (and reallocation), examining the relationship between market mechanisms and government-based approaches. Globally, a market/commodity viewpoint of water has taken hold in many countries, prompted by and largely going beyond the initial western experimentation with water markets. As Howe and Ingram observe, the West continues to search for the appropriate role for markets within government water allocation systems, balancing concerns of economic efficiency with social goals of equity and the protection of public values.

The protection of public values is also a prominent theme permeating Chapters 3 and 4, authored jointly by David Getches and Sarah Van de Wetering. In Chapter 3, the challenge of environmental protection in the West is reviewed, noting the difficulty of recognizing public values within a legal regime based on private property rights and a commodity viewpoint of water. In Chapter 4, the protection of cultural values takes center stage, with a focus on indigenous water rights. In the American West, the judicial recognition of tribal rights has only rarely been translated into real benefits for native peoples, a pattern with, regrettably, many international analogs.

In Chapter 5, Aaron Wolf explores the significance of international and interstate rivers in promoting regional conflict and cooperation. In many parts of the world – especially the Middle East and Central Asia – it is widely feared that transboundary water conflicts will become increasingly common and violent, although a thorough review of history shows that tolerance, more so than active conflict or cooperation, is the most common trend. With water shortages increasingly characterizing the United States/Mexico border area, as well as several interstate regions, the vast international experience with transboundary water management is especially valuable.

Finally in Chapter 6, Lakshman Guruswamy and Dan Tarlock bring our attention back to the American West, and to the underlying management challenge of sustainability. The role of water management in sustainable development remains only partially defined on the global agenda – for example, the seminal report of the World Commission on Environment and Development (Brundtland Commission) devoted less than one of its 383 pages to water. Guruswamy and Tarlock take us considerably further, emphasizing the need for the West to build upon promising regional strategies for integrated watershed management and recasting the nature of water entitlements. Undoubtedly, these are conclusions of great importance and near universal relevance, even if the specific form of implementation varies from nation to nation.

Doug Kenney, Editor

LITERATURE CITED

Case, Pamela and Gregory Alward (1997), *Patterns of Demographic, Economic and Value Change in the Western United States*, report to the Western Water Policy Review Advisory Commission, Denver, CO: The National Technical Information Service.

Census Bureau (1997), *Population Projections: States, 1995–2025*, Current Population Reports, May, US Department of Commerce, Economic and Statistics Administration.

Census Bureau (2001), *Population Change and Distribution: 1990–2000*, Census Bureau 2000 Brief, April, US Department of Commerce, Economic and Statistics Administration.

Frederick, Kenneth D. and Peter H. Gleick (1999), 'Water and global climate change: potential impacts on US water resources', prepared for the Pew Center on Global Climate Change.

Gleick, Peter H. (2002), 'A global perspective on water issues and challenges: new answers for old challenges', paper presented at the conference, 'Allocating and Managing Water for a Sustainable Future: Lessons from Around the World,' Natural Resources Law Center, Boulder, 11 June.

Gleick, Peter H. et al. (2002), *The World's Water: The Biennial Report on Freshwater Resources, 2002–2003*, Washington, DC: Island Press.

World Commission on Dams (2000), *Dams and Development: A New Framework for Decision-Making*, November, London and Sterling, Virginia: Earthscan Publications Ltd.

1. Water policy and cultural exchange: transferring lessons from around the world to the western United States

James L. Wescoat Jr

My interest in international water policies and cultural exchange stems from work as a landscape architect on design projects in Glenwood Canyon (Colorado), Kuwait, and Abu Dhabi. Each project required that knowledge gained in one region be used to address problems in another region. As a graduate student at the University of Chicago, I took courses on ancient irrigation systems in Mesopotamia, Egypt, and Mesoamerica, but the university offered no courses on western water, which was alternately regarded as exotic and mundane. Later, in Colorado I found much in common between water management in the western United States and other parts of the world, and became convinced that geographic comparison of water policy lessons has practical as well as intellectual value (Wescoat, 2002a).

PROBLEM STATEMENT

This chapter asks, 'How can water management lessons be drawn for the western United States from distant places?'[1] Water management information has expanded dramatically over the past century, especially in recent years with expansion of Internet resources, international projects, travel, trade, and education – offering increased opportunities for comparing water systems. When do comparisons lead to practical water management lessons and applications? At first sight, water management comparison has implicit yet largely tenuous links with application. Judging from the large literature of case studies that draw few practical lessons for other places, intellectual curiosity would appear the aim of many water publications (Wescoat, 1994). Some comparative studies strive to draw generalizations from case studies, through controlled analysis of a small number of cases, such as Maass and Anderson's (1978) comparison of irrigation systems in Spain and the United States, or statistical analysis of a large number of cases such as water data tables published in UN documents or the World's Water biennial series (Gleick, 2000). Other comparisons focus on catastrophic water problems, such as high arsenic concentrations in South Asia and the western United States where increased attention in each region elevated concern in the other. Others use comparisons in the opposite way, distinguishing one place from another, sometimes to justify water policy decisions made on grounds of efficiency, equity, liberty, sustainability, or other criteria believed to be less common in other places.

Comparisons become more practical when they focus on water management 'successes' and 'failures' for their potential relevance beyond the places and times where they have been observed. Some comparisons have explicitly sought to draw water policy lessons from other regions for the western United States, though they are rare. The search for lessons for the western United States is rare in part because it entails difficult theoretical, methodological, and cultural challenges; but the main message of this chapter is that they are feasible, if not essential, for the future of western water management.

To grasp the magnitude of the challenge, consider six common arguments against searching for international water management lessons for the western United States:

1. *Irrelevance* – the most extreme argument asserts that each watercourse, water user, and water organization differs from all others, and thus only local solutions have long-term sustainability.[2]
2. *Incompatibility* – this argument asserts that while water users may address comparable problems on comparable watercourses, their solutions cannot be transferred because solutions developed in one context do not fit other contexts.
3. *Incomprehensibility* – this argument suggests that even when potentially compatible water solutions are developed in different areas, people rarely understand them well enough to transfer them successfully from one place to another.
4. *Proximity* – a less extreme view suggests that even if distant solutions are available, compatible, and comprehensible, solutions closer in space and time are more useful than those from distant places and times. As a region develops its own capabilities, it views the search for distant solutions as decreasingly worthwhile. As successful local experiments spread, water managers adopt them without needing formal international comparisons.
5. *Coercion* – this argument takes a radically different tack, arguing that some historical "transfers" are imposed by one place on others in ways that are harmful. Coercion occurs when water treaties are negotiated in

the wake of conquest, water policies extend uncritically across ecosystem boundaries, and risky water management experiments are sited in subject territories. As one Uzbek water manager put it when refusing foreign water management assistance shortly after independence, 'We have our own mustaches!' Anti-globalization movements against privatization of water supplies also strive to resist coercive processes.

6. *The Politics of Difference* – political resistance to international forces is implicit in many of the arguments above. The negative aspects of modern water management transfers, in terms of environmental damage and social impacts, can generate political opposition to outside approaches and trends. Conversely, entrenched local interests can mobilize political resources to maintain the status quo in water management.

Each of these arguments against drawing lessons from one place to another invites elaboration and rebuttal, which could comprise an entire chapter. For present purposes, however, they identify challenges to be addressed when developing arguments about how lessons can be transferred in useful and sustainable ways.

This chapter begins by reviewing changing attitudes toward international lessons in western water management during the 20th century. Following this historical perspective, it examines four approaches for transferring lessons to the West. These are comparative water law, diffusion of water policy innovations, social learning and social movements, and legal transplants. The chapter concludes with a promising path for transferring sustainable water management lessons from around the world to the western United States.

WATER POLICY AND CULTURAL EXCHANGE IN THE 20th CENTURY

The 20th century was an extraordinary period of water resources research, development, degradation, and management – at all scales and in all subsectors (Wescoat and White, 2003). During the course of these developments, the flow of experience between the western United States and other parts of the world changed dramatically.

Early Internationalism in Western Water Management

This story begins in the mid 19th century with a remarkably cosmopolitan attitude among early western water managers. From the 1860s to 1910s there are many examples of active searching for international lessons. In the 1860s, letters were sent to United States consuls and diplomats in Asia, Europe, and the Middle East (Wescoat, 2002c). Congress published their replies in House and Senate documents. George Perkins Marsh (1864, 1874), a diplomat, legislator, and scholar from Vermont wrote about environmental lessons from Europe and the Mediterranean. His widely circulated volume on *Man and Nature: or the Earth as Modified by Human Action* included a chapter on 'waters' and 'forests' that underscored the hazards of watershed degradation. He later submitted testimony to Congress on the 'Evils of Irrigation', urging caution in nascent reclamation programs, again based in part on knowledge of irrigation problems in southern Europe.

In the late 19th century, engineering delegations traveled to Australia, China, Egypt, Europe, and India (for example, Brown, 1904; Davidson, 1875; Hall, 1886; Wilson, 1891, 1894). United States water experts were welcome short-term visitors to water projects in Asia and elsewhere, but they had little detailed knowledge of those systems, in part because colonial powers restricted long-term involvement, with few exceptions, through the 1930s. However, the newly created libraries of land-grant universities purchased foreign publications about irrigation, hydraulics, and agrometeorology, as evidenced in Colorado State University Agricultural Extension reports of the early 1880s. Foreign water specialists – ranging from civil engineers to stonemasons and irrigators – migrated from Asia and Europe to the western United States.

Legal treatises surveyed foreign laws on water and irrigation law (Weil, 1911). Eugene Ware (1905), a lawyer in Topeka, Kansas wrote the first, and only, volume in English devoted entirely to Roman water laws; he also contributed to Kansas's first water lawsuit against Colorado (*Kansas* v. *Colorado*, 1903, 185 US 125; 46 L.Ed. 838; 22 S.Ct. 552). Salt Lake City attorney Clesson S. Kinney's *A Treatise on the Law of Irrigation* (1912) included short chapters on irrigation laws in the Andes, Australia, Canada, China, India, the Ottoman Empire, Egypt, Italy, France, Germany, Great Britain, and Mexico to provide a context for detailed treatment of irrigation laws in the western United States.

As a final example, expansion of 19th century irrigation in places like central Arizona drew upon antecedent Native American canals and water management practices. Spanish missions from California to Texas converted and conscripted Indian cultivators who fused old and new world irrigation practices, laying the foundation for Hispanic and Anglo irrigation. Further east, water management and drainage for rice production along the Carolina coast built upon the transatlantic labor and technological legacy of African slaves (Carney, 2001).

Transition

During the 1910s to 1940s, the search for international lessons slowed down, partly due to financial crises and wars, but also due to the increasing regional capabilities of western water managers. The United States began to export international water expertise and assistance. United States and European flood disaster experts assisted China during the 1920s, under the auspices of the League of Nations (Wescoat, 1995b). United States missionaries had continuing involvement with water and public health programs in European colonies. On a larger scale, the Tennessee Valley Authority began to receive international recognition and emulation in the 1940s, prior to decolonization (for example, in the Damodar Valley Authority in India and Gal Oya Authority in Ceylon). Federal water agencies, such as the United States Geological Survey, Soil Conservation Service, United States Army Corps of Engineers, and United States Bureau of Reclamation also offered limited technical assistance to other nations in the early 20th century. These activities reflect the changing direction of influence between the United States and other parts of the world.

Similar patterns have occurred in other branches of law and policy, where references to Roman law, civil law, and English common law were frequent in the 18th and early 19th century (Wescoat, 2002b). 'The comparative law leavening of indigenous American law largely came to an end by the 20th century due to the influence of the historical school of jurisprudence, the adequacy of West Publishing Company's national reporter and digest system in accumulating a corpus of American law, and the general social force of nationalism' (Clark, 1994, pp. 24–5). Western water law and policy followed a similar trend, one half-century behind the broader national trend.

From Internationalism to Regionalism

The major shift occurred in the post-war era, from the 1950s to 1980s, when western water policy documents ceased to refer to international experience, beginning instead with the constitutional foundations of water management (that is, federal and state responsibilities and relations). For example, the influential President's Water Resources Policy Commission (1950) volumes developed recommendations based explicitly and almost entirely on a foundation of United States laws and experience. The *Federal Reclamation and Related Laws Annotated* (1972) series, which replaced an earlier series of 'green books' on reclamation law, likewise limited themselves to the United States and the 17 western states (the only exceptions being international water treaties to which the United States is a party). Although Federal and state case law cites water laws from Egypt, Peru, England, Italy, India, and

France in the late 19th century, citations narrowed to England, Canada, Spain, and Italy in the early 20th century. In the late 20th century, foreign citations were more likely to focus on failed commercial transactions (for example, breach of irrigation equipment contracts in Iran, Iraq, and Israel) than on water laws.[3]

Interestingly but not surprisingly, the main exception to this pattern involved rivalry with the Soviet Union. Congressional reports assessed Soviet advances in water resources and hydropower development, and scientific translation series made irrigation and hydrologic research in Russian more widely available to United States scientists (US Senate, 1960). Rivalry occasionally gave way to cooperation as in innovative water programs of the International Institute of Applied Systems Analysis in Austria, and the East–West Center in Hawaii. These, again, reflect a broader legal trend in which '… the nineteenth century displayed all the wares and tools available to a comparative legal system … [and] the twentieth century abruptly changed course to turn the American legal system into an inward looking system where legal and rational comparisons are made within the internal legal order controlling in the United States' (Levasseur, 1994, p. 42).

Continuity in an Era of Global Change

The eclipse of international consciousness in western water policy persisted through the late 20th century. For example, the bold yet controversial report on *Water in the West: Challenge for the Next Century*, prepared by the Western Water Policy Review Advisory Commission (1998), referred only to international basins shared with Canada and Mexico, with a passing comment on the Nile River. The leading United States casebooks on water law cite few international water codes or cases (compare, by contrast, the University of Dundee international water law program in Scotland; and the FAO legislative series, Burchi [1994]). A National Research Council (NRC) Report on *New Strategies for America's Watersheds* mentions only a few experiments in the United Kingdom and Australia, neglecting decades of watershed experiments in Asia, Africa and Latin America (NRC, 1999a). Another NRC (1999b) report on large dam management in which I was involved cited a few cases in Canada, Europe, and the World Commission on Dams in South Africa. But compare those few citations with the detailed work of the World Commission on Dams (2000), which produced a wealth of case studies, surveys and country studies.

Important exceptions to this pattern include comparative studies of groundwater management in the western United States and South Asia (Moench, 1991; Moench et al., 1999); United States–Mexico borderlands studies (Ingram et al., 1995); and comparative studies of privatization in

Europe, South America, and the United States (Bakker, 2003; NRC, 2002), to name a few. Comparative historical research has also thrived, expanding its scope to encompass Native American, Chinese, Hispanic, Italian, Japanese, Punjabi, and Scottish irrigators in the West, though the relevance of these academic works for contemporary water policy is unclear (Worster, 1986). In short, western water policy documents seem decreasingly informed about comparable efforts in Africa, Asia, Europe, South America – and their potential relevance for the United States. This trend unfolds even as those regions develop innovative approaches, experiments, capabilities, and knowledge, in part through increased involvement of western water experts.

Increasing Involvement of Western Water Experts in International Projects

Ironically, this inward-looking perspective in western water policy coincided with increasing United States involvement in water development in other regions. Beginning with the first UN conference on Natural Resources (1948) at Lake Success, New York, United States approaches to integrated river basin development started to spread abroad, with United States assistance, though not at home. President Truman's Point Four program established foreign aid for developing countries that, along with pervasive United States influence in the United Nations and World Bank, accelerated the export of United States water expertise and policy approaches.

One export process included training visits for foreign water specialists in the United States, notably at the Bureau of Reclamation headquarters in Denver and at western land-grant universities in Arizona, California, Colorado, Texas, Utah and elsewhere, which received millions of dollars for applied water resources research, education, and training contracts (Freeman et al., 1989; Ives and Bochar, 2002; Wescoat et al., 1992).[4] It is not clear what influence foreign visitors may have had on water managers in the western United States, or what foreign lessons western water specialists brought back to the United States, though those questions warrant detailed study. Nor is it clear what western water managers draw from international and global water forums. Although some United States scholars contribute to the Global Water Partnership, World Water Commission, International Water Resources Association, and other forums, United States water agencies and managers seem less involved than their foreign counterparts. If continued, this trend would mark the decline of United States influence in international water management, which would further limit the prospect of learning from experience in other parts of the world.

Decreasing Clarity about the Nature of Water Policy Exchange

Having shifted from a region that imported water management lessons in the early 20th century to one that exported lessons in the late 20th century, the western United States now finds itself in a new situation, in which an abundance of new knowledge about water management is emerging from different parts of the world, but the interest and capacity to use that knowledge in the western United States remain unclear. In light of this changing situation, it comes as little surprise that there is skepticism about the relevance, compatibility, comprehensibility, or transferability of international water experience for addressing contemporary water problems in the western United States.

While preparing for the conference associated with this book, we had growing enthusiasm about what *might* be learned from distant places and times. Our inspiration stemmed from our shared familiarity with innovative water projects and programs around the world. Examples included applied research on irrigation institutions in South Asia (sponsored by the International Water Management Institute in Sri Lanka), participatory watershed management in India (Farrington et al., 1999), comparative water law in Europe (Seidman and Seidman, 1996; Tarlock, 1997; Wouters, 1997) and in UN agencies (Teclaff, 1972; and the pioneering FAO legislative series of Caponera, 1973, and Burchi, 1991, 1994), and expanding Internet resources such as Oregon State University's Transboundary Freshwater Disputes Database directed by Wolf (2002).[5] All of these activities contributed to a growing collective awareness of how much we might learn from a new internationalism.

Rapid Changes in Water Resources Knowledge

Between 1900 and 2000, the fields of engineering hydrology, hydraulics, aquatic ecology, limnology, hydrography, water law, and other fields of western water policy made extraordinary advances. The growth of Internet water resources information in the past decade alone – ranging from documents to data, models, and decision support systems – defies comprehensive description (see Wescoat and White, 2003, including Appendix A, for an attempt). Information on water management problems and solutions in other parts of the world has become increasingly, though not uniformly, accessible (Wescoat and Halvorson, 2000). In light of this rapid expansion of and access to scientific and policy information, there is no longer any excuse for ignoring international water management experience.

A continuing challenge, however, lies in assessing the comparability of experiments in different environments, spatial scales, and cultural and legal contexts. Our initial inclination was to think, as Montesquieu did in the 18th century, that arid and semi-arid regions might provide the closest analogues to the American West (as, for example, in Grossfeld, 1983; Maass, 1990; Maass and Anderson, 1978; Montesquieu, 1949 reprint; Semple, 1918; Wilhite et al., 1985). But that sort of reasoning can lead to pitfalls of environmental determinism of the sort exemplified in Karl Wittfogel's *Oriental Despotism* (1983/1957) (Wescoat, 2000). Wittfogel erred in many of his comparisons, but he correctly recognized their importance, along with the problems of coercion in water management, which Worster (1986) examined in the western United States.

Using these historical and geographical examples can be understood as processes of cultural exchange. United States laws reflect the application of legal cultures from England, France, Spain, Germany, Italy and, indirectly, Rome (Ware, 1905) to western water problems. They also encompass deeper cultural threads derived from Native American, Mesoamerican, African–American, and Asian–American water management. Add to these the contemporary innovations in watershed management networks in Asia, in irrigation organizations in the Andes, and in new water codes in Africa and the European Union – to name a few – and the opportunities for cultural exchange for addressing western water problems become enormous, if also unwieldy.

Thus, as our conference planning group became increasingly aware of international precedents and analogs that seem promising for the western United States, the central question of *how* to transfer lessons loomed large, and it is to these challenges that we now turn.

TRANSFERRING WATER MANAGEMENT LESSONS: FOUR CONCEPTUAL APPROACHES

The first point to make is that notwithstanding the theoretical, methodological, and historical difficulties noted above, water management innovations continuously move across boundaries. The previous section of this chapter has listed many examples of historical transfers that occur at every scale, from the drip irrigation emitter to complex river basin planning and global climate change research. The charge of this section is to focus on theoretical and methodological approaches that cut across these diverse scales and substantive examples.

We begin with the common sense of water resources comparisons in everyday practice. Common sense comparisons do not by themselves rebut the six counter-arguments against transferring lessons from one place to another, noted at the beginning of this chapter. Thus, this section also examines formal models for understanding cross-cultural exchange including diffusion of innovations models, social learning and social movement theories, and legal transplants.

1. Comparative Theory and Practice

Simple comparisons of water management similarities and differences are common practice and common sense. Irrigators observe their neighbors, as do water consultants, water utilities, water agencies, water activists and, increasingly, international water organizations. Professional organizations like the International Water Resources Association, the International Water Law Association, the Global Water Partnership, the International Water History Association, and the World Water Council, to name a few, facilitate comparison at the international level. Every table, matrix, graph, and map of water data is comparative, as are textual compilations and classifications of water laws (for example, Easterly, 1977; Gleick, 2000; and Radosevich et al., 1976). However, the actual logic, utility, and consequences of water resources comparisons are less well documented.

Because comparisons are commonplace, it seems worthwhile to reflect upon their variety and logic when applied to drawing lessons from international water management experience (cf. Chodosh, 1999; Kahn-Freund, 1966,

Table 1   Drawing lessons of difference and similarity

A. Discriminating differences:
   1. Existential discrimination focuses on presence and absence (that is, of a water management innovation or the necessary conditions for its success).
   2. Differential discrimination distinguishes between degrees and kinds of transfers that are possible (for example, can an innovation be transferred wholesale or in limited increments? Does it need to be adapted in one or many ways?).
   3. Refinement of discrimination through practice (that is, what kinds of expertise are needed to determine whether a water management innovation is transferable?).
B. Making associations across similar cases:
   1. Associations of things (for example, recognizing similar water management practices and conditions that make transfers feasible).
   2. Associations of ideas (for example, conceptual approaches like watershed planning, best management practices, irrigation efficiency, and water budget analysis that facilitate transfer of water management practices based on them).
   3. Associations of ideas and things (for example, through trains of thought that link similar ideas with similar things; that is, B1 and B2).

1974; and Stein, 1997 for similar reflections on comparative law). Recall from the first section of the chapter that comparisons may aim to identify differences that preclude one approach and justify another, or similarities that justify transferring lessons from one region to another. The philosopher William James (1890) and others have shed light on the psychological processes involved in discriminating differences and making associations across cases (Table 1).

James's typology helps us understand the basic situations in which water policy transfers occur, and thus decide on the analytic methods (that is, quantitative, qualitative, or hybrid) needed to evaluate those transfers (for example, Ragin, 1987). The comparisons presented in water data tables, graphs, and maps require more detailed analysis to determine how similarities and differences affect the transfer of lessons across regions. Some logicians are deeply skeptical about judgments of similarity (Goodman, 1972), which are more often imaginative analogies than truly stable similarities. In the water resources field, judgments about the similarities and differences across cases can change dramatically as social groups struggle to redefine concepts, methods, and scales of water management to achieve their political-economic and cultural, rather than logically consistent, aims (Michel, 2000).

In fact, analogy has special relevance for transferring water laws, policies and practices (Glantz, 1988; Meyer et al., 1998; Wescoat and Glantz, 1998). How often do we hear metaphorical references to water as 'lifeblood', 'priceless', or 'sacred'? Frequent analogies – positive and negative – have been drawn between the privatization of water supplies in the United States and events in Bolivia, Chile, England, and France, as well as with earlier eras of public and private water institutions in 19th-century New York, Boston, and Philadelphia (Mentor, 2001). Our colleagues in jurisprudence have much to contribute to our understanding of the logic, uses, and limits of analogies in western water law – and much to learn from the role of analogy in other legal systems, such as Islamic water law where analogy (*kiyas*) is one of the four main branches of legal reasoning (Caponera, 1973; Getches, 1993). For example, the right of thirst for animals in Islamic water law is based on analogies to human thirst, a comparison that has not been but could very well be transferred to United States environmental law (Wescoat, 1995).

Regrettably, international water management comparisons rarely use these sophisticated approaches. They often entail simple juxtapositions of examples, selected unsystematically, from which readers may draw whatever they will, and rarely do so in a way that affects water management practice (for example, Geertz, 1972). Thus, we need to proceed to more formal models for transferring useful lessons from other parts of the world to the western United States.

2. Diffusion of Innovations

Interest in how legal precedents spread was pioneered by Gabriel de Tarde in his *Laws of Imitation* (1903). Diffusion models received more wide-spread attention during agricultural modernization programs of the mid 20th century and most notably in the Green Revolution in Asia and the United States, which involved among other things the adoption of hybrid seeds and associated irrigation, agrochemical, and labor inputs. Close to home, one geography dissertation modeled the diffusion of decisions to use center-pivot irrigation technology on the High Plains of eastern Colorado (Bowden, 1965).

Table 2   Diffusion of innovations models

Categories of those who adopt innovations (for example, irrigation technologies):
   Innovators
   Early adopters
   Early majority
   Late majority
   Laggards

Other actors who facilitate or resist innovation (for example, large dams or micro hydropower)
   Change agents
   Opinion leaders
   Opinion aides

Decision-making processes that lead to the diffusion of innovations
   Awareness
   Persuasion
   Decision to adopt
   Implementation of the decision
   Confirmation (that is, sustained adoption over time)

Social criteria and processes that affect the decision to adopt an innovation
   Relative advantage (is it preferable relative to existing and alternative
      water management practices?)
   Compatibility (will it fit with other social and environmental practices?)
   Complexity (is it simple enough to transfer to a new context?)
   Trialability (can it be tried out on a limited scale?)
   Observability (can the benefits be observed?)
   Reinvention (can it be easily modified to fit new situations?)

What are the key concepts and variables in the diffusion of innovations? Sociologist Everett Rogers (1995), a leader in this field for three decades, outlines the key elements in diffusion models as shown in Table 2.

Diffusion of innovation research emphasizes the role of communication processes, paths, and media in predicting the spread of innovations, such as watershed planning or drip irrigation in recent years. It analyzes the difference that access to information makes, as well as the frequency of contact with different communication media, networks, channels, structures, persons, and messages. Geographers, led by Torsten Hagerstrand's (1967) research group at Lund, Sweden, have focused on spatial diffusion modeling (Brown, 1981; Morrill et al., 1988). Although these formal models have not been applied to international water policy innovations to date, they have reached a level of refinement that would make such analysis possible for innovations both in water law and management.

Diffusion models should not be applied uncritically. They can be biased toward 'adoption' and against 'laggards', and are thus associated with cultural processes of modernization and globalization (Blaut, 1993), though they can focus as readily on the rediscovery and adaptation of traditional technologies. Diffusion research is more descriptive than explanatory – somewhat akin to the type of legal analysis that can be done with automated Shepardizing™, which can describe the path of water law citations while doing little to explain them or evaluate their performance.

3.   Social Learning and Social Movements in Water Management

While diffusion research can help describe and simulate transfers of water management innovations to the western United States (for example, of drip and subsurface irrigation), other lines of social research focus on causal mechanisms and processes. Social learning and social movements are two examples that, while not commonly discussed together, parallel one another in interesting ways for water resources management.

Parson and Clark (1995) provide a survey of social learning theories applicable to Adaptive Ecosystem Assessment and Management (AEAM), including adaptive management of regional water systems. A key principle in adaptive management is that societies can 'learn by doing' ecosystem experiments (Daneke, 1983; Iles, 1996; Lee and Lawrence, 1986; Walters and Holling, 1990). Following their early development in Canada, adaptive management of water systems is now underway in the California Bay-Delta (CALFED), Glen Canyon Dam (and the Grand Canyon Monitoring and Research Center), the Columbia River Basin, the Everglades, and the Upper Mississippi River Basin, to name some of the larger water-related programs in the United States (Jacobs and Wescoat, 2002). Adaptive management has diffused to international projects of the Global Environmental Facility (GEF), for example, Danube and Black Sea water quality programs, which may yield results relevant for United States experiments.

Parson and Clark (1995) discuss five types of social learning theory relevant for water and environmental management:

1. Individual learning in social settings and conditioned by social forces
2. Learning by social groups and organizations
3. Learning in and through science
4. Learning in policy-making
5. Evolutionary theories relevant to learning (that is, on the limits and adaptive consequences of learning)

According to Parson and Clark, the key research questions related to social learning include: Who learns? What do they learn (for example, behaviors, facts, concepts, works, skills, opinions, attitudes, and/or values)? How did they learn? What counts as learning? And why should we ask about social learning? Although this last question may seem rhetorical, it is warranted because so little social scientific research is currently underway on multi-million dollar adaptive management programs in the United States. While millions are spent on ecosystem research and on coordinating stakeholder processes, little research focuses on how stakeholders actually use and act upon ecosystem monitoring data and research results.[6] Similarly, we know almost nothing about how international experience affects decisions by western water managers, notwithstanding the copious information, travel, and international project experience available to them.

An important international initiative on social learning about global environmental issues is underway in the United States and Europe (Social Learning Group, 2001). This initiative examines how organizations and networks learn about ill-defined problems like global climate variability, stratospheric ozone depletion, biodiversity conservation, and acid rain – and how they draw lessons for national, regional, and local policy – all of which are directly analogous to the question of how water managers transfer lessons from other parts of the world. Although global change research is at an early stage of development, it is already clear that scenarios of hydroclimatic variability in the western United States are shaped by the broader field of global change research and, moreover, that they have been transferred from basic research to system operations and policy analysis.

Social movement theories parallel and complement those of social learning. They seek to explain how social groups organize to preserve or change human–environment relationships (for example, the shift from reclamation to anti-dam movements). Social movement theories range from individualistic psychological approaches, in which a disturbance generates anxiety that leads to 'resource mobilization' and collective action (for example, Smelser, 1962; Tilly, 1978), to contextual analyses of 'political opportunity structures' that guide a movement's strategy (Tarrow, 1994); they also cover theories of social ideology, hegemony and revolution (Gramsci, 1971); and the so-called 'new social movements' that are transnational and transcultural in scope (Meleuchi, 1980).

Each of these theoretical approaches may shed light on water management lessons for the West in ways that may seem surprising. For example, psychological anxiety about the water resources implications of global climate change is evident in environmental journalism and public policy, especially in the drought years since 2000. Environmental activists strategize about the political opportunity structures that may enable or constrain mobilization of their resources and constituencies for stream restoration, endangered species protection, and urban water conservation. New social movements have arisen in opposition to both large dams and privatization of water supplies, and they have used a wide array of social, geographical, and cyber spaces to organize themselves. The International Rivers Network (IRN) has supported dam opposition worldwide and has drawn lessons for its United States campaigns. The World Commission on Dams, which included IRN and other dam opponents with hydropower and water engineers, may itself be an example of a new social network that strives to link environmental, social, and economic development; and, as mentioned earlier, it has generated a wealth of new knowledge and lessons for dam operations, monitoring, and decommissioning in the West. There are still other lessons to be learned: revolutionary social change in countries like South Africa, for example, has facilitated some of the most dramatic water law reforms worldwide. Although water movements may be exciting or aggravating, depending on where you stand, social movement theorists rightly argue that such processes should not be dismissed as special interests nor mythologized as utopian visions, but rather used for the practical insights and lessons produced through conflict, struggle, and cooperation (Rangan, 2000; Swyngedouw, 2004).

By focusing on international events that bring about individual and collective change, as compared with the adoption or rejection of a foreign innovation by an otherwise unchanging social group, social learning and social movements research thus complements research on the diffusion of innovations (Miller, 2001). In addition, social learning and movements research sheds light on changes in water organizations, while diffusion research concentrates on water management technologies and legal precedents. To round out this theoretical survey, we need to take a closer look at theoretical perspectives on legal change.

4.   Legal Mirrors and Legal Transplants

For non-water lawyers like me, water law is a fascinating mirror of our culture and its relationships with nature. *Coffin* v. *Left Hand Ditch, Kansas* v. *Colorado, Audubon Society* v. *the Superior Court of Alpine County*: these cases together with hundreds of others and associated water statutes, regulations, and law review articles fascinate me and my students. They help us understand how water users in different parts of the western United States have adapted to varied environmental situations, transformed them, weighed alternatives, struggled for advantage, and adjusted to changing conditions. They are a great cultural archive, as are the water laws of every nation that has faced comparable challenges – some of them, like China, India, and Mesopotamia, for thousands of years compared with our few centuries.

Water laws mirror environment–society relationships. For environmental determinists, like Montesquieu, water laws reflect the influence of nature on culture. Semple (1918) wrote an early treatise on how climate and physiography purportedly shaped water laws in Europe. Although many of these notions have been empirically refuted by local and international examples, it is still common to hear western water laws justified on the basis of climatic conditions. For social constructivists, by contrast, water laws adjust conservatively to changing social wants and values (Bentham, 1962; Brint and Weaver, 1991; Maine, 1888). For law and economics theorists, economic forces drive legal change, while for social movement theorists the primary forces are political. Although they disagree on the logic of social action, utilitarians, pragmatists, and postmodernists agree that societies can make laws and policies that travel across environmental and international boundaries.

These competing arguments about what shapes water laws, and what water laws reflect about environment and society, share the common belief that laws mirror something beyond themselves (Ewald, 1995). At the other end of the spectrum lie so-called theories of 'black-letter law', where legal texts are interpreted more literally and self-referentially. For Alan Watson, a Roman law specialist and comparativist, the most compelling approach lies between these two poles, in what he calls legal transplants (Watson, 1976, 1993; Wescoat, 2002b).

Briefly, a legal transplant is a law drawn from one place and time and applied in another, often with modification. For example in Colorado, a decision in the Division 1 Water Court in Greeley in 2001 might influence a decision in the Division 5 Water Court in Glenwood Springs in 2002. Further afield, attorneys might invoke a Roman *praetor's* interdict to argue a case on the Mississippi River, the Arkansas River, Lake Michigan, Texas, or Mono Lake (Baade, 1990, 1992; Sax, 1971; Wescoat, 1997, 2002b). Proposed transfers may be rejected, as when provisions of the 1873 Canal and Drainage Act in India were explicitly deemed incompatible with legal and social conditions in the San Joaquin Valley of California, also in 1873 (Wescoat, 2002c).[7] Contemporary research on legal transplants relevant for the United States draws more upon developments in international environmental law than municipal law, which involves transfers across jurisdictional scales as well as space (Social Learning Group, 2001; Tarlock, 1997; Wiener, 2001; and Wouters, 1997).

As in this chapter, legal transplant theorists argue, first, that transfers are common; second, that they involve *legal* change, as such, as well as whatever other social and environmental changes they mirror; third, that transplants are directed by legal elites; and fourth, that the reception of legal transplants depends upon several practical factors.[8] Legal transplants are not entirely dependent upon or independent of other social and environmental processes. Instead, legal elites (for example, jurists, judges, and legislators) and elite legal institutions (for example, bar associations and law schools) decide what legal transplants will and will not occur, and what social and environmental processes and arguments will justify them (Watson, 1995).[9] Other comparativists do not deny that laws travel, but they debate the degree of external influence on legal change, and they present alternative arguments about the purpose, efficiency, and social reception of legal transplants (Heim, 1996; Kahn-Freund, 1974; Markesinis, 1990; Mattei, 1994; Reimann and Levasseur, 1998; Stein, 1997; Watson, 1996).

Legal transplant theories have four lines of relevance for transferring water lessons to the western United States. First, contrary to Watson's argument about the frequency of legal transfers, we have observed declining international influence during the 20th century, notwithstanding increasing international sophistication. Second, if useful legal precedents are 'out there', then theoretical and methodological perspectives from research on the diffusion of innovations, social learning and social movements may help explain where and how they might be transplanted. Third, following Watson and Ewald, if legal transplants have their own logic in elite legal institutions, far greater scholarly attention to comparative water law will be needed to effect promising legal transplants in the western United States, which requires elite legal movements.[10] Finally, if legal transplants depend upon elites, special attention should be focused on uneven power relations and coercion among those elites, with special concern for the equity and sustainability of their activities.

CONCLUSIONS AND IMPLICATIONS

The trends and theoretical perspectives examined here indicate that there are many historical precedents and sound theoretical approaches for transferring water management lessons from distant places and times to the western United States. They also shed light on the wealth of international experience that warrants consideration. This chapter highlighted international advances in irrigation management, watershed management, and water law reform – live issues in the West on which a wealth of scientific evidence has developed in other parts of the world and has yet to be tapped.

While we observed declining attention to international experience in many spheres of western water management in the 20th century, which indicates limited awareness, we have also witnessed increasing international involve-ment in ways that could plant the seeds for drawing lessons from other places (Bradlow, 2000; Freeman et al., 1989). With advances in water management around the world, and accessible information about them, the time now seems right for greater cultural exchange of water management solutions.

The theoretical section of this chapter argued that academic curiosity and simple comparison help identify promising transfers but do little to imple-ment them. Research on the diffusion of innovations helps forecast where and how such transfers can most likely occur. Diffusion by itself is insufficient to transfer sustainable water management practices, which requires social learn-ing and movements to understand how and why change will occur. Sustainable transfers also depend upon a deeper understanding of the logic and pitfalls of legal transplants. Sustainable legal transplants, in turn, require more serious academic engagement with comparative water law and policy, which brings us back full circle to the initial role of basic compilations and comparisons. Collectively, these theoretical, methodological, and empirical advances seem capable of tapping the wealth of historical and geographical evidence on international management, identifying promising new transfers, and respond-ing creatively to counter-arguments noted at the beginning of this chapter.

In the end, it seems important to recognize that western water managers are pragmatists, in both the everyday and philosophical senses of that term (Brint and Weaver, 1991). They prefer concrete, useful, familiar solutions to academic research and geographically distant ideas. But they also focus on finding solutions to water management problems, and within reason will cross whatever boundaries or processes of cultural exchange are involved to search for promising solutions to water management problems – from the scale of the individual irrigated plot to the degraded urban stream, complex river basin, and hydrosphere.

NOTES

1. This chapter is dedicated to the Natural Resources Law Center for its leadership in the field of water resources research, policy, and law. As a young doctoral student just starting

to work on water issues in the Upper Colorado River Basin, Center Director Larry MacDonnell and many others patiently helped newcomers comprehend the logic and limitations of western water law. That same spirit has continued with each change of staff and focus. In his characteristically generous way, Doug Kenney has given thoughtful advice on ways to strengthen this chapter. I am honored to be able to offer something back. I am also deeply grateful to Gilbert F. White, mentor for all geographers in the field of water resources, who has showed by example how lessons can travel sustainably through science and policy research.

2. While Gilbert F. White (1957) emphasized that every river is different, he also drew valuable generalizations about international river basin development.
3. This conclusion is based on a LEXUS/NEXUS search for 'irrigation' AND country name; May 2002.
4. Harvard and Cornell were also extensively involved, for example, in Pakistan and Sri Lanka. Their programs had indirect influences on western water policy (for example, through analytic methods at Harvard and irrigation systems and bureaucratic research at Cornell) (for example, Uphoff, 1992). The Ford Foundation's Rural Poverty program supported Cornell's research on irrigation organizations in Asia and later supported related research on water and poverty in the southwestern US (Brown and Ingram, 1987).
5. Transboundary Freshwater Dispute Database, http://terra.geo.orst.edu/users/tfdd/
6. Though see the informal exchange of experience underway in a recently formed Adaptive Management Practitioners' Network – http://www.iatp.org/AEAM
7. The term 'legal transplants' does not appear once in case law of the federal government or that of any of the western states. LEXIS/NEXIS Legal Universe, all courts and dates; consulted 12 May 2002. Cf. Butterworth, 1980, on transplants in oil and gas conveyancing.
8. These vary by author, some stressing simplicity and fairness (Morriss, 2001). Watson (1996) mentions four factors: extreme practical utility, chance, difficulty of clear sight, and need for authority.
9. Elite concerns for status and authority may help explain the more common recourse to Roman and modern European law than to less familiar but also less regarded Islamic, African, and Asian laws.
10. The University of Colorado at Boulder Law School is regarded by some western water managers as having this role.

LITERATURE CITED

Baade, H.W. (1990), 'Springs, creeks, and groundwater in nineteenth-century German Roman-law jurisprudence with a twentieth-century postscript', in David S. Clark (ed.), *Comparative and Private International Law: Essays in Honor of John Henry Merryman*, Berlin: Dunker and Humboldt, pp. 61ff.
Baade, H.W. (1992), 'Roman law in the water, mineral and public land law of the southwestern United States', *The American Journal of Comparative Law*, 40, 865–77.
Bakker, K. (2003), *An Uncooperative Commodity: Privatizing Water in England and Wales*, Oxford: Oxford University Press.
Bentham, Jeremy (1962), 'Of the influence of time and place in matters of legislation', in *The Works of Jeremy Bentham*, vol 1, New York: Russell & Russell, pp. 171–94.
Blaut, James M. (1993), *The Colonizer's Model of the World: Geographical Diffusionism and Eurocentric History*, New York: Guilford.
Bowden, Leonard W. (1965), *Diffusion of the Decision to Irrigate: Simulation of the*

Spread of a New Resource Management Practice in the Colorado Northern High Plains, Chicago: Dept of Geography, University of Chicago.

Bradlow, D. (2000), 'Report on international and comparative water law applicable to large dam construction', prepared for Thematic Review V.4: Regulation, Compliance and Implementation Options, Washington College of Law, USA.

Brint, M. and W. Weaver (eds) (1991), *Pragmatism in Law and Society*, Boulder, CO: Westview Press.

Brown, F.L. and H. Ingram (1987), *Water and Poverty in the Southwest*, Tucson: University of Arizona Press.

Brown, H. (1904), 'Irrigation under British engineers', *Transactions, American Society of Civil Engineers*, 3–31.

Brown, L. (1981), *Innovation Diffusion: A New Perspective*, New York: Methuen.

Burchi, S. (1991), 'Current developments and trends in the law and administration of water resources: a comparative state of the art appraisal', *Journal of Environmental Law*, 3(1), 69–91.

Burchi, S. (1994), *Preparing National Regulations for Water Resources Management: Principles and Practice*, FAO Legislative Study 52, Rome: Food and Agriculture Organization.

Butterworth, Sharon L. (1980), 'A legal transplant as the solution to the insufficient fraction problem in oil and gas conveyancing', *Tulane Law Review*, 54, 1141ff.

Caponera, D. (1973), *Water Laws in Moslem Countries*, 2 v. Irrigation and Drainage Paper 20/1–2, Rome: Food and Agriculture Organization.

Carney, Judith Ann (2001), *Black Rice: The African Origins of Rice Cultivation in the Americas*, Cambridge, MA: Harvard University Press.

Chodosh, H.E. (1999), 'Comparing comparisons: in search of methodology', *Iowa Law Review*, 84, 1025ff.

Clark, D.S. (1994), 'The use of comparative law by American courts (I)', *The American Journal of Comparative Law*, 42, 23–59.

Daneke, Gregory A. (1983), 'An adaptive-learning approach to environmental regulation', *Policy Studies Review*, 3(1), 7ff.

Davidson, G. (1875), 'Irrigation and reclamation of land for agricultural purposes in India, Egypt, Italy, etc.', Executive Doct No. 94, US Senate, 44th Congress, Washington, DC.

Easterly, E.S. III (1977), 'Global patterns of legal systems: notes toward a new geojurisprudence', *The Geographical Review*, 67, 207–20.

Ewald, W. (1995), 'Comparative jurisprudence (II): the logic of legal transplants', *The American Journal of Comparative Law*, 43, 489–510.

Farrington, J., C. Turton and A.J. James (eds) (1999), *Participatory Watershed Development: Challenges for the Twenty-First Century*, Delhi: Oxford University Press.

Freeman, D. et al. (1989), *Local Organizations for Social Development: Concepts and Cases of Irrigation Organization*, Boulder, CO: Westview Press.

Geertz, C. (1972), 'The wet and the dry: traditional irrigation in Bali and Morocco', *Human Ecology*, 1(1), 23–39.

Getches, D.H. (1993), 'From Ashkhabad to Wellton-Mohawk, to Los Angeles: the drought in water policy', *University of Colorado Law Review*, 64, 523–53.

Glantz, M. (1988), *Regional Responses to Global Change: Forecasting by Analogy*, Boulder, CO: Westview Press.

Gleick, P.H. (2000), *The World's Water, 2000–2001*, Washington, DC: Island Press.

Goodman, N. (1972), 'Seven strictures on similarity', in Goodman, N. (ed.), *Problems and Projects*, New York: Bobbs-Merrill, pp. 437–47.

Gramsci, A. (1971), *Selections from the Prison Notebooks*, New York: International.
Grossfeld, Bernhard (1983), 'Comparative law: geography and law', *Michigan Law Review*, 82, 1510 ff.
Hagerstrand, T. (1967), *Innovation Diffusion as a Spatial Process*, Chicago: University of Chicago Press.
Hall, W.H. (1886), *Irrigation Development*, Sacramento: State Office.
Heim, Steven J. (1996), 'Predicting legal transplants: the case of servitudes in the Russian Federation', *Transnational Law and Contemporary Problems*, 6, 187ff.
Iles, A. (1996), 'Adaptive management: making environmental law and policy more dynamic, experimentalist and learning', *Environmental and Planning Law Journal*, 13, 288–308.
Ingram, Helen, Nancy K. Laney and David M. Gillilan (1995), *Divided Waters: Bridging the US–Mexico Border*, Tucson: University of Arizona Press.
Ives, R.H. and R.M. Bochar (2002), *From the Colorado River to the Nile and Beyond: A Century of Reclamation's International Activities*, Boulder, CO: Natural Resources Law Center, University of Colorado School of Law.
Jacobs, J.W. and J.L. Wescoat Jr (2002), 'Managing river resources: lessons from Glen Canyon Dam', *Environment*, 44 (2 March), 8–19.
James, W. (1890), 'Discrimination and comparison', pp. 483–540; and 'Association', pp. 550–604, *The Principles of Psychology*, 1, New York: Dover Publications.
Kahn-Freund, O. (1966), 'Comparative law as an academic subject', *Law Quarterly Review*, 82, 40–61.
Kahn-Freund, O. (1974), 'On the uses and misuses of comparative law', *The Modern Law Review*, 37(1), 1–27.
Kinney, C.S. (1912), *A Treatise on the Law of Irrigation and Water Rights*, 2nd edn, 4 vols., San Francisco: Bender-Moss Co.
Lee, K.N. and J. Lawrence (1986), 'Adaptive management: learning from the Columbia River Basin Fish and Wildlife Program', *Environmental Law*, 16(3), 431–60.
Levasseur, A.A. (1994), 'The use of comparative law by courts (II)', *American Journal of Comparative Law*, 42, 41–59.
Maass, A. (1990), *Water Law and Institutions in the Western United States: Comparisons with Early Developments in California and Australia, Contemporary Developments in Australia, and Recent Legislation Worldwide*, Boulder, CO: Western Water Policy Project, Natural Resources Law Center, University of Colorado School of Law.
Maass, A. and R. Anderson (1978), *...And the Desert Shall Rejoice: Conflict, Growth, and Justice in Arid Environments*, Cambridge, MA: MIT Press.
Maine, Henry Sumner (1888), *Ancient Law; Its Connection with the Early History of Society and its Relation to Modern Ideas*, 3rd American (from 5th London) edn, New York: H. Holt & Company.
Markesinis, B. (1990), 'Comparative law – a subject in search of an audience', *The Modern Law Review*, 53(1), 1–21.
Marsh, George P. (1874), 'Irrigation: its evils, the remedies, and the compensations', US Senate Misc. Doct No. 55, 43rd Congress, Washington, DC.
Mattei, Ugo (1994), 'Efficiency in legal transplants: an essay in comparative law and economics', *International Review of Law and Economics*, 14, 3ff.
Meleuchi, A. (1980), 'The new social movements: a theoretical approach', *Social Science Information*, 17(2), 199–226.
Mentor, Joe, Jr (2001), 'Globalization and water resources management: the changing value of water', paper prepared for the AWRA/IWLRI-University Of Dundee

International Specialty Conference 'Trading Water, Trading Places: Water Marketing in Chile and the Western United States', 6–8 August.

Meyer, W.B. et al. (1998), 'Analogues', in S. Rayner and E. Malone (eds), *Human Choice and Climate Change: Tools for Policy Analysis*, Columbus, OH: Battelle Press, Chapter 4.

Michel, S. (2000), 'Place, power and water pollution in the Californias: a geographical analysis of water quality politics in the Tijuana-San Diego metropolitan region', PhD dissertation, University of Colorado, Boulder.

Miller, Byron (2001), *Geography and Social Movements*, Minneapolis, MN: University of Minnesota Press.

Moench, M. (1991), *Sustainability, Efficiency and Equity in Groundwater Development: Issues in India and Comparisons with the Western US*, Oakland, CA: Pacific Institute.

Moench, M., E. Caspari and A. Dixit (eds) (1999), *Rethinking the Mosaic: Investigations into Local Water Management*, Kathmandu: Nepal Water Conservation Foundation.

Montesquieu, Baron de (1949), *The Spirit of the Laws*, translated by Thomas Nugent, New York: Hafner Press.

Morrill, R., G.L. Gaile and G.I. Thrall (1988), *Spatial Diffusion*, Newbury Park: Sage.

Morriss, Andrew P. (2001), 'Lessons from the development of western water law for emerging water markets: common law vs. central planning', *Oregon Law Review*, 80, 861ff.

National Research Council (NRC) (1999a), *New Strategies for America's Watersheds*, Washington, DC: National Academy Press.

National Research Council (NRC) (1999b), *Downstream: Adaptive Management in Grand Canyon and the Colorado River Ecosystem*, Washington, DC: National Academy Press.

Parson, Edward A. and William C. Clark (1995), 'Sustainable development as social learning: theoretical perspectives and practical challenges for the design of a research program', in *Barriers and Bridges*, New York: Columbia University Press, pp. 428–60.

President's Water Resources Policy Commission (1950), *A Water Policy For the American People*, Washington, DC: Government Printing Office.

Radosevich, G.E., V. Boira, D. Daines, G. Skogerboe and E. Vlachos (eds) (1976), *International Conference on Global Water Law Systems*, 4 vols, proceedings of conference in Valencia, Spain, Fort Collins: Colorado State University.

Ragin, C.C. (1987), *The Comparative Method: Moving Beyond Qualitative and Quantitative Strategies*, Berkeley, CA: University of California Press.

Rangan, H. (2000), *Of Myth and Movements*, London: Verso.

Reimann, Mathias and Alain Levasseur (1998), 'Comparative law and legal history in the United States', *American Journal of Comparative Law*, 46, 1–15.

Rogers, E. (1995), *Diffusion of Innovations*, 4th edn, New York: The Free Press.

Sax, J. (1971), 'The public trust: a new charter of environmental rights', in *Defending the Environment*, New York: Alfred Knopf, pp. 158–74.

Seidman, Ann and Robert B. Seidman (1996), 'Drafting legislation for development: lessons from a Chinese project', *American Journal of Comparative Law*, 44, 1ff.

Semple, E.C. (1918), 'The influences of geographic environment on law, state, and society', in A. Kocurek and J.H. Wigmore (eds), *Formative Influences of Legal Development*, Boston: Little, Brown, pp. 215–33.

Smelser, N. (1962), *Theory of Collective Behavior*, New York: The Free Press.

Social Learning Group (2001), *Learning to Manage Global Environmental Risks*, 2 vols, Cambridge, MA: MIT Press.

Stein, E. (1997), 'Uses, misuses – and nonuses of comparative law', *Northwestern University Law Review*, 42, 198ff.

Swyngedouw, E. (2004). *Social Power and the Urbanization of Water: Flows of Power*, Oxford: Oxford University Press.

Tarde, Gabriel de (1903), *The laws of Imitation*, translated from the 2nd French edn, by Elsie Clews Parson, New York: H. Holt & Company.

Tarlock, A. Dan. (1997), 'The influence of international environmental law on US pollution control law', *Vermont Law Review*, 21, 759ff.

Tarrow, S. (1994), *Power in Movement, Social Movements, Collective Action, and Politics*, New York: Cambridge University Press.

Teclaff, L.A. (1972), *Abstraction and Use of Water: A Comparison of Legal Regimes*, Department of Economic and Social Affairs, ST/ECA/154, New York: United Nations.

Tilly, C. (1978), *From Mobilization to Revolution*, Reading, MA: Addison-Wesley.

Uphoff, Norman. (1992), *Learning from Gal Oya: Possibilities for Participatory Development and Post-Newtonian Social Science*, Ithaca, NY: Cornell University Press.

US Department of the Interior (1972), *Federal Reclamation and Related Laws Annotated*, vol I through 1942, vol II 1943–58, vol III 1959–66, Washington, DC: US Government Printing Office.

US Senate, 85th Congress, 2nd Session, Committees on Interior and Insular Affairs and Public Works (1960), *Relative Water and Power Resources Development in the USSR and the USA*, report and staff studies pursuant to Senate Resolution 248, 4 January, Washington, DC: Government Printing Office.

Walters, C. and C.S. Holling (1990), 'Large-scale management experiments and learning by doing', *Ecology*, 71, 2060–8.

Ware, Eugene F. (1905), *Roman Water Law: Translation from the Pandects of Justinian*, St Paul, MN: West Publishing Co.

Watson, Alan (1976), 'Legal transplants and law reform', *Law Quarterly Review*, 92, 79ff.

Watson, Alan (1993), *Legal Transplants: An Approach to Comparative Law*, 2nd edn, Charlottesville, VA: University Press of Virginia.

Watson, Alan (1995), 'From legal transplants to legal formants', *American Journal of Comparative Law*, 43(L), 469ff.

Watson, Alan (1996), 'Aspects of reception of law', *American Journal of Comparative Law*, 44, 335ff.

Weil, S.C. (1911), *Water Rights in the Western States*, vol 1, San Francisco: Bancroft-Whitney Co.

Wescoat Jr, J.L. (1994), 'Varieties of geographic comparison in the Earth Transformed', *Annals of the Association of American Geographers*, 84, 721–5.

Wescoat Jr, J.L. (1995), 'The "right of thirst" for animals in Islamic water law: a comparative approach', *Environment and Planning D: Society and Space*, 13, 637–54.

Wescoat Jr, J.L. (1995b), 'Main currents in multilateral water agreements: a historical-geographic perspective, 1648–1948', *Colorado Journal of International Environmental Law and Policy*, 7, 39–74.

Wescoat Jr, J.L. (1997), 'Toward a modern map of Roman water law', *Urban Geography*, 18, 100–5.

Wescoat Jr, J.L. (2000) 'Wittfogel East And West: changing perspectives on water development in South Asia and the US, 1670–2000', in A.B. Murphy and D.L. Johnson (eds), *Cultural Encounters with the Environment: Enduring and Evolving Geographic Themes*, Totowa: Rowman & Littlefield, pp. 109–32.

Wescoat Jr, J.L. (2002a). '"Beneath which rivers flow": water, geographic imagination, and sustainable landscape design', in U. Fratino, A. Petrillo, A. Petruccioli, and M. Stella (eds), *Landscapes of Water: History, Innovation, and Sustainable Design*, Bari: Uniongrafica Corcelli Editrice, vol. 1, pp. 13–34.

Wescoat Jr, J.L. (2002b), 'The landscapes of Roman water law', *Environmental Design: Journal of the Islamic Environmental Design Research Centre*, special issue on multicultural Mediterranean landscapes, 88–99.

Wescoat Jr, J.L. (2002c), 'Water rights in South Asia and the United States: comparative perspectives, 1873–1996', in John F. Richards (ed.), *Land, Property and the Environment*, Oakland: ICS.

Wescoat Jr, J.L. and M. Glantz (1998), 'Aral region water issues in comparative perspectives: what can we learn from analogues?', Social Science Research Council, Aral Sea Basin Workshop proceedings, 19–21 May, Tashkent.

Wescoat Jr, James L. and Sarah Halvorson (2000), 'Ex post evaluation of dams and related water projects: patterns, problems and promise', report to the World Commission on Dams, South Africa.

Wescoat Jr, J.L., Roger Smith and D. Schaad (1992), 'Visits to the US Bureau of Reclamation from South Asia and the Middle East, 1946–1990: an indicator of changing politics and programs', *Irrigation and Drainage Systems*, 6, 55–67.

Wescoat Jr, J.L. and G.F. White (2003), *Water for Life: Water Management and Environmental Policy*, with contributions from Mark Ferrara on 'Appendix A: Guide to Internet Resources', Cambridge: Cambridge University Press.

Western Water Policy Review Advisory Commission (1998), *Water in the West*, Denver: WWPRAC.

White, G.F. (1957), 'A perspective of river basin development', *Law and Contemporary Problems*, 22(2), 157–84.

Wiener, J.B. (2001), 'Something borrowed for something blue: legal transplants and the evolution of global environmental law', *Ecology Law Quarterly*, 27, 1295ff.

Wilhite, D.A., N.J. Rosenberg and M.H. Glantz (1985), 'Government response to drought in the United States: lessons from the mid-1970s', from *Drought Response in the United States and Australia, A Comparative Analysis*, Lincoln: University of Nebraska, Center for Agricultural Meteorology and Climatology.

Wilson, Herbert M. (1891), 'Irrigation in India', USGS 12th Annual Report, 1890–91, part II, Irrigation, Washington, DC.

Wilson, Herbert M. (1894), 'American and Indian irrigation works', *The Irrigation Age*, 107–9.

Wittfogel, K.A. (1981), *Oriental Despotism: A Comparative Study of Total Power*, 2nd edn, New York: Vintage Press.

Wolf, A.T. (ed.) (2002), *Atlas of International Freshwater Agreements*, Nairobi: United Nations Environment Programme.

World Commission on Dams (2000), *Dams and Development: A New Framework for Decision-Making*, London: Earthscan.

Worster, D. (1986), *Rivers of Empire: Water, Aridity, and the Growth of the American West*, New York: Pantheon Books.

Wouters, Patricia (ed.) (1997), *International Water Law: Selected Writings of Professor Charles B. Bourne*, Boston: Kluwer Law International.

2. Roles for the public and private sectors in water allocation: lessons from around the world

Charles W. Howe and Helen Ingram

With assistance from Don J. Blackmore, Murray–Darling River Basin Commission; Miguel Ricardo Solanes, UN Economic Commission for Latin America and the Caribbean; Joachim Blatter, University of Konstantz; and Marcus Moench, Institute for Social and Environmental Transition

Among the most salient features of water management regimes are the mechanisms and rules regarding water allocation. In most situations, the institutional mechanisms employed utilize both public-sector and private-sector roles. In this chapter, we review the key issues associated with various approaches. In part 1, the public/private sector issue is described in detail, with an emphasis on the experience in the western United States. In part 2, our focus broadens to consider experiences in other nations, with examples from Australia, Latin America, Europe and Asia. Finally in part 3, this wealth of experience and experimentation is used to identify general lessons, including conclusions and recommendations specific to the western United States.

1. SEEKING THE PUBLIC SECTOR/PRIVATE SECTOR BALANCE

Public and Private Responsibilities

For most of the world, the water management question is not one of complete centralized public sector planning or complete privatization of the resource and its distribution but the proper assignment of responsibilities to each sector within an adequate social oversight framework. In this context, the public sector must be interpreted broadly to include agencies at state and federal levels, river basin commissions, state courts and engineer offices, publicly sanctioned water trading organizations (water banks), special districts and even citizen-initiated watershed groups. The private sector itself consists of a vast range of organizations, large and small, that specialize in a range of tasks from simple logistic functions to the organization of water markets, consulting services, and operation and ownership of some urban water utilities. Most countries are struggling with this balance as globalization presents a new environment in which private actors from around the world are energetically seeking greater roles in the water area.

All systems for the allocation of water are nested within some regulatory structure established through a political/legal process. Rules within basins may be different than those that apply between basins, and still different than those between political regions. Surplus waters may be subject to market allocations, while subsistence levels may be governed by political allocations. Crisis situations may merit yet another set of rules, for example emergency water banks, expropriation of water for priority uses and endangered species protection.

At one extreme, Hawaii opted in 1987 for water allocation through a Commission on Water Resources Management that has complete responsibility for the assignment and transfer of water rights (Gopalakrishnan, 2002; Moncur, 1989). At the other extreme, Chile opted in 1981 to privatize all bulk water resources in an attempt to stimulate an active market in water rights and to greatly reduce State administration (Bauer, 1998). In the United States, various forms of privatization of urban water services are receiving increased attention (National Research Council, 2002), while the marketing of bulk water supplies is increasing in importance not only in the western United States but in the eastern parts of the country where the riparian doctrine long held sway.

To what extent can expanded roles for the private sector, including markets and market-like processes, be beneficially utilized within a politically determined institutional framework? Insofar as water markets are used, how are public values not reflected in private transactions to be protected? These are certainly key questions surrounding water reallocation as shown by Ingram's study of water farms (Oggins and Ingram, 1989) and Howe's work on Arkansas Valley transfers (Howe et al., 1990; Howe, 2000).

Water markets take many forms, from informal rural water sales in countries like Nepal and India (Moench, 2002) to highly organized market arrangements in semi-arid parts of the developed world (Blackmore and Ballard, 2002). Prominent examples of organized water markets include 'capacity-sharing' arrangements on the Murray-Darling system in Australia under the Murray-Darling Commission, similar arrangements on the Snake River in Idaho under the management of the Bureau of Reclamation, and the California water bank that was established under severe drought conditions in 1991

(MacDonnell et al., 1994). The Northern Colorado Water Conservancy District supports a very active market that more closely approaches the many buyer–many seller structure of competitive markets in which the District plays only the role of bookkeeper (Michelsen, 1994).

Water Allocation Tasks and Public/Private Capabilities

The pervasive attribute of water today is *scarcity*, measured by the marginal social value (monetary or otherwise) of water in its various uses. Scarcity is mirrored in the high economic and environmental costs of developing new, reliable supplies. Combined with growing demands for water, these high costs imply that the *reallocation* of existing supplies is of increasing importance. The question then is, what mechanisms should be used to effect reallocations of existing supplies?

Water supplies are allocated, formally or informally, over space, across classes of users and over time. Allocation can be accomplished by a wide range of institutional arrangements, ranging from the 'rule of capture' (still the effective policy for groundwater in many parts of the world) to permit systems and private property arrangements based on riparian or priority doctrines. These are questions that nations everywhere are dealing with. As noted above, some opt for systems that emphasize markets; some opt for systems that emphasize strict policies and major roles for public agencies; most utilize a mixture.

Allocation of bulk water among uses and sectors

Three broad categories of water use are generally distinguished: agriculture and related activities; municipal and industrial uses (M&I), including residential, commercial and public uses; and environmental uses that include biodiversity, water quality, the aesthetics of streams and water-based recreation. Competition among these growing uses in the face of nearly fixed supplies is becoming more intense over time, especially in the semi-arid regions of the world. Non-commensurable cultural and economic values greatly complicate water allocation decisions.

Irrigated agriculture is the predominant diverter and consumer of water in semi-arid and arid regions. In the western United States, irrigated agriculture accounts for about 80 per cent of all surface water and groundwater diversions and about 90 per cent of all consumptive use (USGS, 1998). Similar proportions are found in many regions (Gleick 2001), stemming from historical patterns of development in which agriculture is often the first industry to be established. Inappropriate regional location of agriculture, inefficient irrigation techniques and wasteful cropping patterns result from policies of water and crop subsidies that fail to confront the water-using region and the individual user with the correct scarcity value (or 'opportunity cost') of the water and the real social value of the crops being produced. These features of irrigated agriculture suggest agriculture as a major potential source of water for the other sectors.

Naturally, not all of the water being used in irrigated agriculture is suitable for or available to other uses for reasons of distance, cost, quality and reliability. There is widespread fear that, because 'water runs uphill towards money', M&I demands will dry up irrigated agriculture. That fear is partially justified at the local level where large transfers significantly reduce local irrigated acreage, accompanied by secondary effects on the local economy. On the other hand, a small percentage reduction in selected agricultural water consumption can provide for the doubling of M&I uses. In the longer term, western irrigation agriculture may reverse the post-1902 United States pattern with relocation of large parts of irrigated field crops to more appropriate regions.

Transfers of water within the agricultural sector and between agriculture and other uses have a long history in the western United States under the 'appropriations doctrine' of water law. In Colorado, water rights have been traded as personal property for more than 100 years through contacts between individual buyers and sellers. In Idaho on the Snake River, individuals having storage space in Bureau of Reclamation reservoirs have leased their stored water to others since the 1920s. In the Northern Colorado Water Conservancy District that became operational in 1957, permanent sales and temporary rentals of water have shifted heavily in favor of M&I users.

Market-like arrangements for agricultural to M&I transfers are found in Southern California where the Metropolitan Water District and San Diego County have contracted to improve the irrigation system of the Imperial Irrigation District with resultant water savings diverted to the District and County. Other market-like arrangements include *drought year lease-outs* that have been arranged between the Metropolitan Water District and the Palo Verde and Desert Irrigation Districts (MacDonnell et al., 1994).

Municipal and industrial water users usually have a greater ability to pay (higher prices) for water than the agricultural sector. M&I users also require high quality and reliability (Howe and Smith, 1994). Outdoor uses account for more than 50 per cent of M&I withdrawals (AWWA, 1992). Changes in landscaping practices (xeriscaping) can significantly reduce total consumption and would help avoid summer peak demands.

M&I supplies are typically metered to individual users so that volumetric pricing can be applied. Increasingly, urban utilities are using increasing block rate structures to reflect the high marginal cost of providing water and to motivate conservation. It is well known that M&I demands are responsive to price (significant price elasticities), but this responsiveness decreases with increases in the incomes of households (Renwick and Archibald, 1997). In spite of the more frequent use of increasing block rate structures, M&I water is typically underpriced because public water agencies, while accounting for infrastructure costs, often fail to recognize the scarcity value of the raw water itself. The Boulder, Colorado water utility uses an increasing block rate structure keyed to drought conditions but attributes no cost to existing raw water supplies that may have a market value of $300 to $500 per acre-foot. If revenues covered this cost, it would add at least $1.00 per thousand gallons to the current rates.

Water-based recreation is important in many countries. Studies in the western United States have shown that participants value these activities quite highly and are willing to pay substantial amounts for streamflow maintenance, higher water quality and protection of riparian ecosystems (National Research Council, 2002). The newest water demand in high-income regions is for environmental purposes (as discussed in Chapter 3). Increasing incomes and educational levels lead to demands for higher-quality environments and to concern with issues like biodiversity. The provision of these water-based environmental amenities is mostly dependent on public agencies because the benefits are widespread and cannot be individually rationed or priced (such services are called 'public goods'). For the same reason, significant new instream flows are required by laws and regulations like the United States Endangered Species Act and the Clean Water Act that have required the reallocation of stream flows from traditional uses to instream flows. A few cities and agricultural interests have dedicated some of their water rights to instream flow protection, including Boulder, Colorado that dedicated water rights having a market value of $14 million to late season instream flow. These values, along with community and cultural values tied to water, emphasize the need to broaden the legal concept of 'beneficial use' and to broaden the 'no injury' rules that currently constrain water transfers in the West.

Water also stands as a symbol of cultural values and particular lifestyles. 'Sparkling mountain spring water' as a kind of trope for Rocky Mountain living fails to be included in the usual economic analyses of water benefits. Water relates to culture and identity and carries a heavy emotional content. Water has become a kind of 'quality of life product' as it has been for thousands of years in the ancient Mughal water gardens of Northern India (Wescoat, 1989).

As society's changing values create pressure for reallocation of existing water supplies, the mechanisms for reallocation must be decided on – an often painful process. Just as new economic and non-economic values must be acknowledged, so should the old values that have to give way. This is an area where efficiency (shifting towards the newer uses) may clash with equity toward those who hold the old values. Traditionalist 'water buffaloes' resist new uses and new institutions even when it may be in their long-term economic interests to surrender to the change. This is where voluntary transactions through water markets can play an important role by generating price information and offering higher prices to the traditional users. Environmental interests can buy water rights for these new uses (as recommended by Colorado Supreme Court Justice Greg Hobbs, 2002). Having to enter the market for water for, say, environmental and recreational purposes is a real test of value and is preferable in the minds of many to applications of the 'public trust doctrine' and other mandated forms of water reallocation.

Allocation among regions
The river basin is the natural physical unit for water management. Naturally, interbasin transfers of water are widely used but the physical connectedness of the river basin makes it the natural management entity. Major problems are likely to arise when the political subdivisions that have policy-making powers over water resources fail to correspond to the boundaries of the basin. This disjunction of basin and political boundaries results in 'jurisdictional externalities', that is, losses of basin-wide benefits through failure to recognize gains or losses to other political jurisdictions. For example, the Colorado River is divided into Upper and Lower Basin jurisdictions that stem from the 1922 Colorado River Compact. The distribution of available water among the four states of the Upper Basin and, in turn, the three states of the Lower Basin is determined by compact and court orders. In the absence of interstate water markets, no state has motivation to recognize the value of additional water to the other states.

In the United States, the federal government has ceded much of the regulatory power over water resources to the individual states. The McCarren Amendment (43 USC 666, 1954) requires federal claims to water to be adjudicated in state courts. This is also true in Canada, Australia, and Brazil. Each of the states of the western United States has its own system of administering property rights over water, although most states abide by the 'appropriations doctrine'. One of the consequences is that, to date, water markets and water transfers have been confined to the individual states, that is, there are no interstate water markets even though some of the greatest inefficiencies in allocation occur between states. In 1991, California proposed an interstate marketing arrangement for temporary (annual) exchanges of water among states with the state water authorities acting as marketing organizers (State of California, 1991). The idea was quickly shot down by Colorado, as was the State Engineer who had the temerity of endorsing the concept without clearing it with the Governor. Other well-conceived schemes for interstate leasing of foregone agricultural consumptive uses – like the

Resource Conservation Group scheme of 1990 – have received no serious
political consideration (Lochhead, 2001).

This may be changing. The Bureau of Reclamation has been working with
the Lower Basin states of California, Arizona and Nevada to identify beneficial trading patterns among those states. Nevada and Arizona are currently
negotiating an arrangement in which Nevada would purchase annually some
of Arizona's excess supply from the Central Arizona Project (CAP) that
Arizona would then recharge into their aquifers for future use. When Nevada
needs the water, they would divert it from the large Colorado River reservoir,
Lake Mead, and Arizona would, at the same time, pump the same amount as
a substitute for taking water into the CAP (*U.S. Water News*, May 2002,
p. 17). This would build upon existing water bank arrangements between the
two states.

The complexity of international river and groundwater allocation exceeds
that of interstate allocation. Famous historical cases include the Nile, the
Ganges, the Danube and Rhine Rivers and, in the United States, the Colorado
and Rio Grande Rivers. The general problem is that the upper-basin state has
physical control of the water. Unless another issue can be negotiated simultaneously, the lower-basin state has little leverage. The compacts negotiated
between Mexico and the United States over the Colorado and Rio Grande
present an interesting case of simultaneous negotiation in which the United
States was the upper-basin state on the Colorado while Mexico was, to an
important extent, the upper-basin state on the Rio Grande.

In considering public and private arrangements for inter-regional and international allocation of water, the public sector must play the major role in
facilitating 'win-win' agreements. Within a public policy framework that
protects public values, markets may have a role as was proposed by California for heavily supervised interstate trades on the Colorado River.

Allocation over time

Different issues are involved in allocating renewable and non-renewable
water supplies. Allocating renewable supplies usually involves a short time
horizon, over the annual water cycle or over several year cycles of wetness
and drought. Sustainability of the renewable systems is the critical issue.

Non-renewable supplies present more difficult decisions of a 'now versus
future' nature over much longer time horizons. The 'social discount rate'
plays a major role (implicitly or explicitly) in such cases. Large non-renewable water resources may determine the economic and demographic fates of
regions like the non-renewable groundwater in the Ogallala aquifer in the
High Plains region (Nebraska to Texas) of the United States. Mapp (1972)
demonstrated that pumping from the Ogallala was faster than economically
optimal on the shallow parts of the aquifer (intertemporal externalities),

while pumping closely corresponded to an economically optimal pattern on deep parts of the aquifer. Some type of regulation and the use of economic instruments (taxes) would be desirable in the former situation.

In contrast to the non-renewable groundwater case, the Edwards aquifer in Texas (Kaiser and Phillips, 1998; Keplinger and McCarl, 1996) is a large regional karst formation that is readily recharged by precipitation and quickly drawn down by agriculture, urban pumping and the discharge from several highly valued large springs that support endangered species. This cycle runs from one to several years depending on climate conditions. The large alluvial aquifer of the South Platte River in Colorado (used mostly for agriculture) exhibits a six to eight year cycle of draw-down and recharge dependent on wet and dry climate patterns (MacDonnell, 1988). Some aquifers are artificially recharged when surplus water is available.

Obligations to future generations and biodiversity complicate questions of allocation over long periods of time and require criteria that go beyond standard benefit–cost approaches. Given the short time horizons typically used by the private sector and the long-term implications of sustaining renewable systems and judiciously using non-renewable resources, it is clear that the public sector must play a major role in setting regulations and using economic instruments like appropriate prices, penalties and taxes to motivate practices that are consistent with the long-term viewpoint. Precautionary principles should apply to irreversible commitments of resources, while humans should respect the rights of other species to continue to exist and flourish (Costanza, 1991).

Social Objectives of Water Management: Comparative Advantages of Public/Private Approaches

The major objectives sought by society through the allocation of resources (not just water) can be classified as (1) economic efficiency, (2) equity and (3) sustainability. The concepts may be difficult to define clearly, but it is even more difficult to determine to what extent they might be complementary rather than competitive. We will concentrate on how the achievement of each might be affected by differing public and private roles in water allocation.

Economic efficiency
Economic efficiency refers to an allocation of scarce resources that maximizes the difference between benefits generated and cost incurred. The concept can be applied to macro-level decisions (like designing agricultural or water policies) or to individual project design and selection. At the latter level, benefit–cost analysis (B/C) is the major quantitative tool used by public agencies (although similar methods of analysis are used in the private sector).

B/C analysis is required in the United States for all water projects that involve federal participation (Howe, 1988). B/C requires the monetization of the beneficial and detrimental impacts of a project and their summarization as the 'present value of net benefits' that takes into account the time patterns of impacts through the 'discounting' of future values. Two questions follow: 'Can all significant impacts of a project be monetized?' and 'What is the appropriate way of discounting future values?' Suffice it to say that techniques for monetizing positive and negative impacts have been greatly expanded (for example, recreational benefits from reservoir use, valuation of improved water quality and so on) but some impacts of importance to society are considered by many as beyond monetization (for example, preservation of unique ecological systems, cultural values, and the psychological costs of change). Arguments also continue over the appropriate discount rate if not discounting itself. Thus benefit–cost analysis is extremely useful in weeding out projects that are economically bad, but it should not and is not the sole criterion for decision making.

Privatization and reliance on markets are usually touted as conducive to achieving an economically efficient allocation of scarce resources, selection of projects, their designs and operating procedures. Private markets work well when there is competition in the provision of goods, when private transactions affect only the buyer and seller, and when the resultant costs and benefits occur over a relatively short time horizon. The question is whether or not the allocation of water among uses and the selection of water project investments meet these conditions. The answer is obviously 'No' for several reasons: a change in the uses of water will affect water users other than the buyer and seller (through return flow effects); important benefits and costs of water use and water projects don't result in money revenues; there usually are few buyers and sellers; and because most water projects are not attractive to most private investors because of the long payback period.

Thus water allocation cannot be left entirely to markets and the private sector. The privatization of water services must take place in a regulatory framework that protects or recognizes all socially relevant benefits and costs that occur over the relevant long time horizons.

Equity
Equity is a catch-all term that 'exists in the eye of the beholder' but still serves as a label for important social values that are tied to the *distribution* of benefits and costs among impacted groups. It includes issues of non-monetizable community and cultural values not likely to be taken into account by private market participants. In the early days of B/C analysis following World War II, economic efficiency was promoted as *the* criterion to be used in project design and selection, that maximizing the 'size of the pie' was what

counted and that equity issues would be settled through various social safety net and tax programs (for example, Musgrave, 1959).

It is now recognized that equity and economic efficiency are dependent on one another for several reasons. First, the prices used in B/C analyses depend on the distribution of income and wealth: a society of landlords and peasants will exhibit a different set of prices than a society with roughly the same resources but consisting of middle-class farmers and professionals. Thus if the distribution of income and wealth is unsatisfactory, B/C analyses will have little social meaning.

Secondly, 'Who gets what' is politically and socially important in itself and the safety nets referred to by Musgrave frequently are non-existent. Thus society is unlikely to assign the same weights to benefits and costs accruing to different groups. 'Efficiency' depends on these social weights.

Finally, failure to address equity issues in project design and selection can impact the stream of benefits and costs if, for example, the pattern of water allocation or project design leads to social unrest or social rejection of the project. Examples are found in water development projects such as the Sardar Sarovar project in the Narmada Valley of India and the Kousou Dam in Ivory Coast where the planned net benefits have been offset by the uncounted costs of social strife and resettlement.

There are also important issues of 'who gets to decide'. Even apart from who wins and who loses, there is a political interest in who is 'at the table' and what 'rules' govern decision making. During the decades of federal dominance of water resources development in the western United States, bureaucratic experts often dominated allocation decisions that might have been made very differently had they been left to grassroots interests. Today, communities are insisting on open and transparent decision making, with scientific information related to water quantity, quality, endangered species, and restoration plans shared with the public.

One public survey of water users in areas of potential loss or receipt of water found that most people agreed that areas of origin suffer losses of control, security, economic opportunity, lifestyle and cultural heritage (Oggins and Ingram, 1989). Majority rule that favors urban areas may not be viewed as equitable. Academic policy analysts and water planners have responded to the numerous equity challenges by developing a variety of multi-objective decision procedures (for example, Major and Lenton, 1979) but such formal procedures are seldom used in practice.

Thus it is clear that commonly held concepts of equity are unlikely to be achieved by private transactors in free markets. This does not imply a public sector monopoly in the provision of water services, but rather the need for a social regulatory framework that will motivate or require private actors to take broader social values into account.

Sustainability

The Brundtland Report (World Commission on Environment and Development, 1987) defined sustainable development as 'meeting the needs of the present without compromising the ability of future generations to meet their own needs'. This is hardly an operational definition, but it imparts the right spirit of the sustainability notion. Sustainability in part relates to the supply side, that is, the ability of a society to maintain the outputs of vital economic and environmental goods and services over time. However, the level of demands for goods and services drives the level of outputs so that sustainability depends on the intensity of our consumption patterns. The achievement of sustainability is likely the more humble our interpretation of 'needs' (Howe, 1997).

The sustainability literature has distinguished 'hard sustainability' from 'soft sustainability', the former indicating that each renewable system should be maintained in a non-impaired state. Soft sustainability emphasizes that the productivity of a high-level system can be maintained while subsystems may come and go, one substituting for the demise of another, akin to the sustainability of ecosystems wherein higher-level systems can be sustainable in the face of fluctuations in their component species (Costanza, 1991). The 'soft' interpretation may be the more relevant one.

Sustainability has a geographical dimension also akin to ecosystems: a larger region may exhibit sustainable production capacities while its subregions do not. This is illustrated by agriculture at the regional, county and farm levels. Sustainability also has a temporal dimension, exemplified by the conifer forests of the Upper Midwestern United States that were extensively cut at the turn of the century, yet have regrown to become valuable recreational areas (Barlowe, 1983). From a long time perspective, this unplanned sequence of events might be judged to be sustainable.

How does sustainability apply specifically to water systems? Sustainability of renewable water systems obviously depends on climatic conditions and protecting the run-off conditions of the watershed or river basin. Given a stable climate regime, catchment areas must be protected from erosion to sustain run-off and storage capacity. A widely ignored issue is the ever-increasing sedimentation of reservoirs. While reservoirs are designed with 'excess' capacity to allow for sedimentation, their capacities are declining over time.

Non-renewable groundwater deposits cannot, by definition, be sustained unless we leave them untapped – usually an uneconomic alternative. However, the larger production system that uses the water may be sustainable if increasingly efficient methods of water application and more responsive crop varieties are developed. This is what is happening in the High Plains region of the United States that is dependent on the largely non-recharging Ogallala aquifer. While some small areas are running out of water, more efficient irrigation and increasingly productive dry-land crops are offsetting most of these losses at the regional level.

It seems clear that sustainability requires social regulation for all the reasons cited earlier in connection with efficiency and equity. Also critical for sustainability are the different time horizons used by the private sector and those needed for the long-term maintenance of productivity of natural resource systems.

Major Unsolved Problems in Water Management and Public/Private Capabilities

Water institutions, both public and private, embody vestiges of historical values and technologies and frequently fail to evolve sufficiently to serve present needs. For example, the attributes of water rights and the definitions of 'beneficial use' and 'no injury' have seriously lagged rapidly changing public values. Institutional innovations have exhibited flaws that were not anticipated by their designers, such as the failure of repayment for Bureau of Reclamation projects and the ecological consequences of large-scale federal water projects. This section will illustrate a few of these problems. These examples are symptomatic of larger problems of institutional fragmentation that occur among levels of government and in relation to the many quasi-public entities (such as conservancy districts) that are involved in the provision of water.

Continued separation of water quantity and quality management

Those concerned with water quality in the early 1970s were understandably suspicious of longstanding state and federal agencies historically in charge of water supply. At the insistence of environmental groups, the regulation of water quality was given to the new federal Environmental Protection Agency and, at state levels, to state boards of health or similar agencies, for example the California Water Quality Control Board. As a consequence, water quality and quantity management are frequently handled separately even though the two are physically inseparable. As a result, interagency conflicts abound and inefficiencies get built into the regulatory structure.

It has been shown that the abandonment of irrigated acreage in shale areas of the Upper Colorado Basin would be less costly in achieving reductions in river salinity than the point source controls and renovation of distribution systems promoted by the traditional agencies. Water quality and public health agencies in California have resisted the increase in the proportion of reclaimed water allowed in municipal water supplies while water supply agencies have encouraged recycling of water. Water quality agencies often regulate the parts per billion of salinity, heavy metals, and pesticides in water, while water allocation agencies determine the amounts of water available for dilution. Separation of quantity and quality regulations is especially problematic when it comes to non-point sources of pollution and their control under the new 'total maximum daily load' (TMDL) standards. Here, effective policies require close collaboration among the agencies dealing with land use, water quality, water supply and agriculture. New multiparty participatory mechanisms that include the involvement of different agencies with jurisdiction over both quality and quantity, such as CALFED and the Everglades Restoration Project, may mitigate some of these difficulties.

Missed opportunities for surface water-groundwater conjunctive management

Because groundwater science and technologies developed more slowly than surface water hydrology and hydraulics, different legal regimes emerged – another failure to integrate law and physical science. In Texas, owners of most surface lands have unlimited rights to pump groundwater in spite of negative externalities exerted on neighboring pumpers and future users. Unless the price of energy for pumping becomes prohibitive, there is little incentive for groundwater users to take these externalities into account or to use more costly renewable surface water supplies. As a result, every user pumps too much and costs rise too rapidly. Arizona has placed agricultural groundwater use under a regulatory regime in which applications per acre are supposed to be reduced over time until overdraft of aquifers ceases to occur. Similar reductions in per capita use apply to urban water use, but with no cap on total urban use.

There are many potential benefits to integrating groundwater and surface water management together in so-called 'conjunctive use' approaches. Aquifers provide natural storage with very low evaporative loss. The transmission of water through the aquifer to many users requires no infrastructure. Long-term storage to meet severe droughts is provided. Nonetheless, conjunctive management is practiced in relatively few areas of the West. Highly organized conjunctive management is found on the South Platte River in Colorado and on the Edwards aquifer in Texas. Conjunctive management has long been used in Southern California where groundwater replenishment districts have been set up to enable municipal water utilities to balance surface and groundwaters according to their availabilities. These replenishment districts also buy excess surface water from regional water suppliers such as the Metropolitan Water District at low rates to recharge aquifers (Blomquist, 1992).

Lack of congruence of hydrologic and administrative boundaries

As mentioned earlier, the failure of political and administrative boundaries to match hydrologic regimes continues to bedevil water management. In fact, jurisdictional externalities may have become a more serious problem with the fading influence of large federal agencies that had some basin-wide authority. Interstate sales of water have not been accepted politically even where they make enormous economic and environmental sense. The 'severe sustained drought study' described by the Powell Consortium (1995) showed that the current institutional framework for the Colorado River is not appropriate for periods of sustained drought because of inflexibilities in interstate allocation. While California, the Department of Interior and several private parties have proposed arrangements for interstate leases on the Colorado River, the Upper Basin states have opposed the idea for fear of permanent loss of their water rights (Lochhead, 2001). More viable are *intrastate* reallocations; Vaux and Howitt (1984) demonstrated that intrastate reallocations within California could generate large net benefits.

Water districts that receive water from federal and state projects generally confine water sales and leases inside of district boundaries on the grounds of protecting the repayment base for project costs (Wahl, 1989), although such trades in California have been facilitated by the 1992 Central Valley Project Improvement Act that explicitly allows districts contracting with the federal Central Valley Project to permit out-of-district sales. Even when inter-jurisdictional sales and leases represent win–win possibilities, acceptance problems often persist because of the inability of winners in one jurisdiction to compensate losers in another (Howe, 2000).

Where multiparty collaborative arrangements such as CALFED in California are convened in good faith, some jurisdictional externalities can be overcome. In Brazil, the national water agency, Agencia Nacional de Aguas, is working with the riparian states of several river basins to design basin-wide planning procedures (ANA, 2001). On a smaller scale, currently popular local watershed associations are dealing with many local watershed issues, although the source of the problems they seek to solve may lie outside the individual watershed, and their own actions may affect others 'downstream'. There have been no studies of the aggregate consequences of programs proposed by individual watershed groups on the basin as a whole. Consequently, watershed associations may generate the same jurisdictional difficulties that previously led localities to turn to the federal government for more comprehensive solutions.

Roles for the Private Sector in Water Allocation and Management

The potential for private sector activity in the development and provision of water services received increasing attention during the 1990s. This international interest is part of a global quest for improving the availability of water and wastewater services, service reliability, cost reduction, water quality, technological resources, and the timely maintenance and replacement of aging water infrastructure. Smaller water systems face the greatest difficulties in meeting these challenges. For smaller systems or communities, the 'regionalization' or consolidation of management and operations across several communities can achieve scale economies, extension of service areas and performance improvements.

The term 'privatization' covers a wide spectrum of water utility operations, management, and ownership arrangements, ranging from the simple provision of supplies and auxiliary services through operation and maintenance of urban and rural systems to private ownership of the physical infrastructure (National Research Council, 2002). The contracting of management and operations to a private provider has been more common than the sale of utility assets. In the United States, no major city has sold its utility assets in recent decades, although the practice is found more frequently among small utilities (National Research Council, 2002). Because of variations in political, demographic, economic, and physical circumstances, no single model of private water service responsibilities best fits all situations. Both public and private water and wastewater management are motivated by several similar objectives, including proper maintenance of infrastructure and protection of the basic water source. On the other hand, public authorities may be more responsive to water user needs and social and cultural values. While privately owned and operated water systems are structured to produce profits and may be better motivated to reduce operating costs. Private operators may be less tied to local politics and may have greater flexibility to make staffing changes than publicly owned systems. *A major impact of the increasing competition from the private sector has been to stimulate improved performance in publicly owned and operated water agencies.*

The role for water markets
World-wide recognition of the importance of environmental protection, combined with rapid population growth, has increased the competition between traditional uses of water and newly emerging environmental uses. These developments emphasize the importance of being able to transfer raw water supplies from older and lower-valued uses (especially agriculture) to new emerging uses. Water markets, while long existing in the several parts of the world, are playing an increasing role in effecting these transfers. With appropriate social oversight to protect public, environmental and social values, water markets can be effective.

In designing mechanisms for water allocation, the following criteria should be considered (Howe, Schurmeier and Shaw, 1986b): (1) flexibility in allocation over time; (2) security of tenure for water owners; (3) reflection to the water user of the real opportunity cost of the water being used; (4) fairness to the participants in the water system; and (5) allowance for water users to adjust the levels of risk they face from hydrologic uncertainty. Proponents argue that water markets fulfill these criteria fairly well. Since market transactions are on a 'willing seller–willing buyer' basis, the transaction is apparently 'fair' to both buyer and seller, while the water owner has security of the property right on which long-term plans can be laid. Where active water markets exist, the water user is continually confronted with market prices that reflect the actual opportunity costs. Even where traditional water prices are distorted by subsidies and outdated repayment agreements, the market price will show the real cost of water use. Finally, where an active market exists, water users know they can always go into the market to secure additional water to meet growth or drought protection needs.

Various types of water markets and market-like arrangements have been used (MacDonnell and Rice, 1994). Water rights in Colorado have been traded among users on a permanent basis for more than a century even though no centralized trading arrangements have been in place. Individual buyers and sellers scouted around for satisfactory 'deals,' facilitated by the legal interpretation of water rights as personal property subject to exchange while subject to water court oversight (MacDonnell, 1989). The efficient water market in shares of the Northern Colorado Water Conservancy District has been extensively described and analyzed (Howe, Schurmeier and Shaw, 1986a; Tyler, 1992). The operations of the Idaho and California water banks are described in MacDonnell (1994), Wahl (1989) and MacDonnell and Rice (1994). The State of New Mexico has just established a water market on the Pecos River to facilitate efficiency-increasing trades among users (*Albuquerque Tribune*, 5 March 2002), while Colorado is in the process of designing a water market in the Arkansas Valley to facilitate temporary trades with the hope of preventing further out-of-basin sales. 'Market-like arrangements' would include the drought year lease-out arrangements between the Metropolitan Water District (MWD) of Southern California and the Palo Verde and Desert Irrigation Districts, as well as the agreements between MWD, San Diego County and the Imperial Irrigation District to invest in improvements in the irrigation system, with the salvaged water going to MWD and the County.

Concerns about Privatization and Water Markets

Some of the important values affected by water systems share two characteristics: (1) the benefits can be enjoyed by many people without diminishing the benefit for others; and (2) it is usually impractical to require people to pay for the benefit. An improvement in water quality is an example for many downstream parties can enjoy it while it would be difficult to make them pay for this benefit. Private firms may ignore community and cultural values. Such a benefit is called a 'public good'. Public goods are significant because private parties tend not to provide them adequately if at all. The issue then arises as to how to protect such values under privatization.

Privatization is not equivalent to competition; that is, long-term contracts between the private sector and public utility lack the discipline of ongoing competition that occurs only at contract renewal time. Continuing oversight and monitoring arrangements are important. Although private operation and management of water services may provide savings in operating budgets and capital costs, rates charged to customers usually must rise as contractors undertake the promised upgrading of the system while generating a profit. The prospect of higher rates has been a deterrent to privatization.

Concerns about privatization have arisen in part from some unfortunate experiences (while some successes have been overlooked). The most famous Third World case of attempted privatization of town water supply is the case of Cochabamba, Bolivia. The city's water system, operated by the municipal water company Semapa, was in poor condition and didn't serve poor sections of the city, so many had to buy water from vendors who charged more than the company. The World Bank had pushed for privatization while the Bolivian government pushed for inclusion of some expensive features against the Bank's advice. A contract was signed with an international consortium to operate and expand the city system. With the high-cost items to be covered, the consortium increased rates an average of 35 per cent, with low-income areas paying about 10 per cent more than before and high-income areas paying more than double the previous rate. At least 50 per cent of the increases appear to have been due to the expensive add-ons. Protests followed and escalated into rioting with one death and many injuries. The state government finally rescinded the contract with the consortium pursuing legal remedies. The case illustrates the emotional environment in which water matters are decided, the uncertainty of outcomes and the risks to all privatization participants.

Public values are unlikely to be taken into account by private participants in the market for raw water. In the water resources area, these values include social and cultural values, instream values, external costs imposed directly on other parties due to jurisdictional boundaries that relieve water users of liability for damage, and 'secondary economic impacts' imposed on areas-of-origin. These values are particularly important in traditional, low-income communities. In the southwestern United States, *the old Spanish acequia systems* not only support local agricultural needs, but maintain social cohesion because maintenance of the canals and distribution of the water are community efforts. Brown and Ingram (1987) have studied the relationships between access to water and poverty in the Southwest.

Cultural values associated with water are not confined to particular groups. Farm families place a high value on the farm or ranch lifestyle. Kenneth Weber interviewed farmers engaged in agriculture in the Arkansas Valley of Colorado who 'stick it out' on marginally profitable farms because they value the farm lifestyle (Weber 1989a, b; 1990). Even after selling the water from their lands, many farmers retain their farm homes. Weber found that of 36 Crowley County, Colorado farmers who had sold their water, 34 remained in the county. This is not to argue that traditional societies should forever remain unchanged, but that the economic 'playing field' is uneven between low-income traditional societies and the more advanced sectors, and that maintenance of these cultures is of concern to society at large.

Balancing Public and Private Roles: The Western US Experience

The current private/public debate in the western United States is based on a heritage of past policies and decisions that have resulted in the following attributes:

- confusing and contradictory legacy of water rights and water law that recognize both the prior appropriation rights of individual users and the water-dependent public welfare of communities;
- great variety of water agencies that have evolved over time, creating fragmentation and conflict but, at the same time creating flexibility and permeability so that parties holding conflicting values have continuing access to decision making;
- legacy of independent grassroots water entrepreneurship that has co-existed with a strong federal emphasis on centralized planning, construction, and management;
- legacy of state government involvement in water market arrangements that has had both positive and negative effects on the potential of this mechanism.

Early history

Much of water law, custom and practice in the American West was inherited from Spain and its four hundred-year conquests and occupation of territory in the New World (Maass and Anderson, 1978). The dual legacy is a marriage of private ownership and public responsibility. While water was considered a public resource to be developed for the public, private ownership and development historically was the mechanism through which the public interest was served. Water rights were allocated along with grants of land, and the grantee had free rein in the use of that water so long as general purposes of public welfare were served. The old Spanish judiciary insisted that water rights be utilized in ways consistent with protecting water quality supplies to towns and villages and insuring that some water was reserved for native tribes (Meyer, 1996).

As the western territories and states joined the Union, each brought somewhat different traditions reflecting different settlement patterns, varying historical experience, and levels of water scarcity. State constitutions institutionalized different state customs and practices. Some state constitutions, like that of New Mexico, specifically stated that water was a public resource. Other states recognized a hierarchy of water uses with municipal uses taking priority over agriculture and mining. Most states recognized the doctrine of prior appropriation (first in time is first in right) while California embraced a unique combination of prior appropriation and riparian doctrines due to its mixed Spanish–English history. Differences among state water laws continue to this day, further complicating the management of interstate waters. States vary in terms of the extent to which they embrace and regulate water markets.

The variation of state water law is particularly pronounced as it affects groundwater. The availability and use of groundwater is varied among western states. Water law was made in state courts and judges were slow to recognize the hydrologic interconnection of surface and groundwater laws. The registration of wells and measurements of use are relatively recent phenomena.

Water-users associations were responsible for much of the early development of water resources in the American West. Dryland farming of many crops is impossible in much of the West, and agriculture necessarily involved the installation of dams and ditches that required capital and cooperation. Some commentators trace the development of civil society in various regions of the West to the comity among neighbors forged in the construction, operation and maintenance of rural water delivery systems (Crawford, 1988). Political leadership training often began through recruitment to the office of ditch rider and membership on local water commissions. While water users at first were regulated by informal norms that grew up over time, the strain of scarcity and drought often caused conflicts that required court settlements and ultimately state legislative action. Along with state courts, the office of state engineer became important in many states as the repository of water rights information and an arbiter of disagreements among rights holders.

Private land and water companies were important entrepreneurs fostering early agricultural settlements. Such companies raised the capital for dam, diversion, and ditch construction through the sale of shares to early settlers from whom they also collected user fees. Where private land and water companies were the instigators of settlement, water districts were organized from the top down and water users were treated more as clients than partici-pating members. Local and state governments enforced the contracts with development companies, and were frequently called upon for rescue when private operations got into trouble financially or were unable to deliver water.

Western cities depended at first upon licensed monopolies through long-term contracts with private water companies. Up until the mid nineteenth century, the City of Los Angeles depended for its water supply on a series of franchises to such companies, with generally disappointing results. Not only was delivery of water falling behind growth, but maintenance of infrastruc-ture lagged. The City withdrew from one failed arrangement after another until it established its own municipal water utility. The municipal utility was able to act more aggressively than previous private suppliers. As early as the 1870s, the City of Los Angeles laid claim to the total supply of the Los Angeles River. The City declared war on upstream users, and won a series of court victories. What could not be achieved through the courts was won by an aggressive campaign of annexation. Expanding the city's boundaries was seen as a way of justifying – indeed requiring – more water to build an even more magnificent metropolis. The annals of water history in the West are filled with stories similar to that of Los Angeles Water and Power. Publicly owned municipal water utilities such as the Denver Water Board and Tucson Water served not just as water suppliers, but as aggressive participants in contests to acquire water rights so that western cities, far into the future, would be assured that cheap, plentiful water would be available for growth.

To promote orderly development of water resources, the federal govern-ment has been a continuing presence in water development. Early in the nation's history, the Army Corps of Engineers established itself as the engi-neering arm of the government through the development of water transportation and flood control projects (Maass, 1957). In 1902, the Bureau of Reclamation was established to build multipurpose water projects throughout the West. The era of large-scale water development in the American West in the middle of the 20th century was dominated by federal agencies, often at war with one another.

Federal roles
Some scholars read the era of large-scale federal water development as runa-way politics in which political entrepreneurs sought to build political capital at the expense of economically efficient developments (Anderson and Leal, 2001). Our understanding differs, and we believe that government involve-ment had both positive and negative consequences. The federal government got strongly involved in water planning and management because it had a vision and the authority and resources to establish a west-wide program of development. It took upon itself several key tasks described below.

The first task was to settle and legitimate water rights. The prime example is the Colorado River Compact of 1922 which divided water between the upper and lower basins as negotiated by the basin states with the strong encouragement of the federal government. Absent the federal involvement, long, expensive and contentious litigation would have been the only alternative. Even with the 1922 compact, continued contention between Arizona and California precipitated one of the longest and most hard-fought court battles in history. Such litigation would have increased exponentially without federal government involvement. Similarly, the action of state engineers and intrastate stream commissions has been essential to resolving instate water rights disputes.

A second task was to undertake comprehensive water planning and development. From the earliest days of the republic, the Army Corps of Engineers built facilities to accommodate navigation to ports and harbors and on interstate rivers. While the Corps was highly responsive to parochial interests, its broad jurisdiction insured a more comprehensive perspective than would have been present in private development. Similarly, the surveys of the Corps of Engineers and Bureau of Reclamation of virtually all the major rivers in the nation provided basic basin-wide geographic and hydrologic information critical to all future development. The differences among state laws and the divisions represented by state boundaries would have made such comprehensive basin-wide perspectives impossible for state governments and private developers. At the same time, the identification of future dam sites on rivers all over the country built up shelves full of plans that could be pressed forward at the first political opportunity. To some extent, such planning represented solutions in search of problems to address, and inevitable droughts and floods were quickly responded to by government agencies with standby construction plans.

A third task was to undertake multipurpose water development. The mission of the Bureau of Reclamation, established in 1902, was to build multipurpose projects that served not just irrigation, but also flood control, electric power needs, fisheries and wildlife, municipal and industrial water supply, and the needs of native peoples. The Flood Control Act of 1936 required the federal government to pay all costs of flood control reservoirs and to cost-share the construction of levees, distorting the motivation that local authorities had to consider the best combination of steps. Storage projects and levees encouraged construction in flood plains that increased flood losses.

Nonetheless, it seems likely that, absent federal agency involvement, particularly through the Bureau of Reclamation, less comprehensive, single-purpose projects would have been pursued, forgoing the possibilities of benefits to a wider public. Benefits to fish, wildlife and native peoples would have gotten even shorter shrift than they received at the hands of the federal government. Private water developments ignored fish and wildlife values and encroached shamelessly on Native American water rights, the latter finally leading to the 1912 'Winters Doctrine' that assured high priority to native water claims.

A fourth task was to apply the disciplines of efficiency and safety. The early 'gospel of efficiency' (Hays, 1959) of the Theodore Roosevelt era referred to 'scientific management' and specialization of land use for the best-suited activities. It wasn't until 1936 (again with the Flood Control Act) that economic efficiency in the benefit–cost sense became a major criterion. The measurement of benefits and costs adopted first by the Army Corps of Engineers and later embraced by all federal water agencies imposed a certain amount of economic, market-like discipline on project construction. Unfortunately, the measurement of benefits and costs was often distorted to bolster political objectives (Howe et al., 1969). Narrow interests were often served by federal projects. Federal subsidies in terms of low interest rates and long repayment periods artificially inflated the demand for projects. Much later, cost-sharing rules were established to impose greater discipline on the choice of projects. The pay-out of huge multipurpose projects like Hoover Dam on the Colorado and the Bonneville complex on the Columbia extended over such a long period that private investment would not have been forthcoming.

A final task of federal involvement in water development was to address equity and environmental issues unlikely to be of concern to private developers. Economically disadvantaged interests have generally fared poorly in private markets. Critics of federal water projects have argued that they have been damaging to both native peoples and the environment (Ingram, 1990). Native American water needs were underserved, and federal overbuilding encroached upon reserved rights Indians had in water. While the trust obligation of the US Department of Interior to protect the water rights of indigenous people has been performed poorly (Brown and Ingram, 1987), the federal involvement provided a structured opportunity for mobilization and protest. Opportunities for public input have been provided by various Congressional acts, including the Fish and Wildlife Coordination Act, the Administrative Procedures Act and the National Environmental Policy Act combined with relatively open authorization and appropriation processes for water projects. While fewer projects would have been built absent federal involvement, private projects would not have provided the structured opportunity to raise equity and environmental concerns available in federal programs.

Modern trends

In the current scene, federal agency involvement in water development has declined and changed in nature since the mid 1960s. The percentage of the federal budget and the amount of congressional time spent in hearings related to federal water management have declined precipitously despite, and perhaps because of, the increased level of criticism of federal water programs beginning in the 1970s. While the proportion of the federal budget spent on public lands and water management hovered between 2.5 per cent and 3.25 per cent from the late 1950s through the early 1970s, since then there has been a precipitous drop, and less than 1 per cent of the budget is currently spent on these areas. The fall in congressional attention has been as dramatic. While in 1948 Congress spent about 18 per cent of its hearing days on these matters, currently it spends only about 9 per cent (Baumgartner, 2002). There is a consensus among critics and supporters alike that the era of large-scale federal water development is a thing of the past.

While the federal government continues to be involved in water decisions, the nature of that involvement has changed. The dictates of the Endangered Species Act, if followed, would halt further public and private development in the habitat areas of threatened species. Rather than strictly and inflexibly applying the law, the United States Department of the Interior has encouraged mediation, and negotiation among affected interests and commitment to habitat improvement (Doremus, 2001). An 'adaptive management' approach has been embraced by federal agencies, particularly in large restoration projects like CALFED and the Everglades Restoration Program. Adaptive management involves creating strong feedback from researchers to managers in the effort to learn how some of the environmental damage of past water projects can be undone (Doyle and Jodrey, 2002). In these cases the Department of Interior leadership provided the impetus for cooperation among different levels of government, and a wide array of private actors including environmental, development, and Native Americans. In these arrangements, substantial water is reallocated to restore wetlands and revive endangered fish populations and scientific experimentation with different management tools is encouraged. Rather than advancing development, the federal government now facilitates conflict resolution and environmental restoration.

Community conservation efforts and watershed associations, frequently with federal and state agency collaboration and encouragement, are sprouting in what was previously thought to be very inhospitable soil. The Trout Creek Working Group in Oregon is one example. The ranchers in the high desert grasslands wanted their livelihood protected and the environmentalists wanted the native cutthroat trout preserved. Contending parties got together and agreed that the watershed and its streams had been so degraded that both ranching and wildlife were threatened. Sufficient mutual understanding made possible a voluntary moratorium on grazing and the creation of a habitat recovery plan (Cortner and Moote, 1999). While it is far too early to judge such efforts successful, grassroots collaborative arrangements represent an important institutional change in water management.

Water marketing and market-like arrangements are experiencing a resurgence although markets often encounter substantial resistance. Water marketing has a 100-year history in Colorado (Rice and MacDonnell, 1993). The Northern Colorado Water Conservancy District (NCWCD) supports an especially efficient water market with nearly continuous trading and low transactions costs. The market uses water from the Colorado-Big Thompson Project, completed by the Bureau of Reclamation in 1957, to deliver water acquired from the western slope of the Rocky Mountains to the NCWCD on the eastern slope. The NCWCD then delivers water to users on the basis of shares owned in the NCWCD. These shares are easily tradable in an active market, and the NCWCD facilitates trades by maintaining a bulletin board of offers to buy and sell (Howe, Schurmeier and Shaw, 1986a). Temporary transfers are also seen. During the severe drought between 1986 and 1991, California and the Bureau of Reclamation operated a water bank. Transfers were for one year only, 1991, during which approximately 800 000 acre-feet changed hands (Howe, 2000).

2. PUBLIC/PRIVATE LESSONS FROM AROUND THE WORLD

James Wescoat (Chapter 1) has noted that differences in context impede the transferability of lessons from one place to another. The very different roles played by government and private sectors in various nations around the globe and in cross-border regions suggest that diffusion of innovative institutions ought to be accompanied with a warning label declaring that results will vary considerably depending upon circumstances. The section that follows will first provide a summary of four case studies from different parts of the world and will draw lessons from each concerning public and private roles in water allocation.

The Murray–Darling River Basin in Australia: a System in Transition

Australia has long been a leader in innovative water administration for reasons that are obvious: water is scarce and a key to agriculture; the region encompasses highly varied natural areas; and it depends extensively on tourism (Blackmore and Ballard, 2002). The structure of governance in Australia is similar to that found in Canada and the United States: strong state governments with certain interstate and international matters reserved for the federal government. The states have had strong power over the administration of water, so the problem of reconciling state and federal policies over river basins is still attracting attention. It is a system in transition to a greater (but not yet clearly defined) reliance on water markets (Young et al., 2000).

The Murray–Darling river system is the largest in the country, extending across three states and one territory while supporting 70 per cent of the irrigated acreage. The Murray–Darling Basin Commission was established to improve and coordinate the administration of water in this complex system. Water trading has been a part of the Murray–Darling Basin environment for the last 20 years. It has been managed on a state-by-state basis with regional trading zones identified by the physical ability to transfer water from one point to another. Each trading zone has progressively become more sophisticated with the availability of better market information and market mechanisms.

To date, the Murray–Darling Basin Commission has focused on expanding the trading zones so that the ownership of water can pass from one state to another. A pilot program of interstate water trading has been in place for three years (Blackmore and Ballard, 2002). The further development of trade is being driven by both the need to facilitate adjustment within the irrigation industry and the requirement to establish a competent and comprehensive market structure prior to the redistribution of water resources to the environment that has been mandated by law.

The following issues need to be addressed to enable the expansion of the current market (Blackmore and Ballard, 2002):

- improved definition of water property rights, including clear separation of water rights from land ownership;
- consistent assessment by the participating states of environmental impacts, especially salinity;
- establishment of procedures to compensate areas that water sales have left with underused delivery and storage systems – so-called 'stranded' delivery assets;
- establishment of protocols to enable transfer of existing rights that have different tenures (length of the life of the right) and reliabilities (akin to the priority of western US rights);
- improvement of administration processes and accounting of trades.

Many groups see socioeconomic disadvantages from water leaving 'their' district. However, there are opportunities to use trading to encourage rural readjustment. Water will tend to trade out of inappropriate areas (for example, salinized or poor soils, flood hazards) as is already happening, but reinvestment in non-water industries may work if other regional attributes are favorable.

Trading is seen as the best way to encourage water to move to its highest use. It is most effective when markets have as few barriers as possible. Acquiring water through government purchases is inevitable if increasing instream flows is the objective. It is better to start earlier than later, as market prices will increase as water moves to higher uses and becomes scarcer.

There needs to be provision for governments to acquire entitlements (rights), compulsorily if necessary, for environmental purposes and to allow closing of parts of the supply system that have excessive delivery losses and operating costs. It is clear that not all of the water needed for increased instream flows can be derived from increased delivery efficiencies. The only other source is water that is already being used for consumptive purposes.

Temporary trades (that is, within one irrigation season) between states have been possible since about 1995 in the River Murray system. The pilot project for permanent interstate water trade was set up in 1997/98. For simplicity it was restricted to the three states of the Lower Murray (New South Wales, Victoria and South Australia) and to high-security (high-priority) entitlements so that trades would involve rights with the same attributes. The first interstate trade took place in September 1998. By September 2000 there had been 51 trades involving 9.8 GL (gigalitres) worth approximately $10 million. 90 per cent of the water moved to South Australia from the upper states.

After two years a review of the project was conducted by CSIRO Land and Water (Young et al., 2000). It found that:

- the approach has been to 'learn by doing' with a spirit of adaptation;
- attention should be paid to improving intrastate trade (as interstate trade has been less than 1 per cent of the total water applied, and a relatively small part of total trade);
- administrative systems need to be streamlined and harmonized. At present trade documents can spend 32 days in the mail, moving from one location to another;
- interstate trade is increasing the value of water use in the Basin. Virtually all trade was from non-users to high-value developments using state-of-the-art technology;
- in the first two years there were no measurable adverse social impacts for the districts that traded water interstate;
- the environmental flow impact has probably been positive, but trade is so small in flow terms that it is impossible to measure;
- salinity impacts are potentially negative, and all states need to improve mechanisms for enforcement;
- buyers and sellers in the market do not understand 'exchange rates' well (an exchange rate is the ratio of the amount of water allowed to be used in the new use to the amount allowed in the original use. The need for exchange rates arises from the need to protect other water users and

water quality). If they are to be used as a mechanism, it will be important to justify and communicate the numbers being used;

● pricing and charging mechanisms across the Basin are inconsistent.

There exist several key barriers to trade. The first is transaction costs. In principle, trade will only occur when the buyer is prepared to pay the price demanded by the seller plus all transaction costs that include legal fees, the costs of technical studies to demonstrate compliance with salinity and environmental criteria, fees to water agencies and government departments, and so on. Transaction costs tend to decrease with increasing market activity and maturity, but will remain significant because water trade is a complex process if the above safeguards are to be maintained.

The second barrier consists of return flow problems. Changing the location of an extraction will increase or decrease instream flows between the two locations. The effects can be beneficial or adverse, and management to minimize the adverse effects increases transaction costs.

Stranded assets constitute a third type of problem: equity concerns. As water is traded out of a region, the fixed costs of supply are spread over a smaller volume, resulting in higher charges to the remaining water users. The problem can be addressed by charging a buyer an 'exit fee' to cover the capitalised fixed supply cost attributed to the water traded out.

Lack of motivation to participate in the water market can also be a barrier to trade. Landholders may be reluctant to sell water because they believe that it will increase in price (for example, it is clear in the western United States that farmers are holding on to water as their retirement investment). There is some evidence that the marginal value of water on the farm in the Murray–Darling Basin tends to be higher than its market value (quite contrary to the western United States where market values of water rights are much above the marginal value in irrigated agriculture). Some landholders perceive the market to be basically 'unfair' while tax considerations favor short-term leases.

The Privatization of Water in Latin America: Water Rights, the Beneficial Use Doctrine and Regulatory Failure

Most water laws have provisions that require the effective use of water entitlements. The principle of effective and beneficial use is widespread. While the terminology is not uniform, in the German law (as amended on 23 September 1986), the 1985 Spanish law, the Mexican water law (art. 27. III), the legislation of most Argentinean provinces, and the laws of the states of the American West, there is a notion that water rights risk forfeiture if not used, or if not used according to the terms of a license or permit (Solanes, 2002).

In the United States, a typical statement of the rule of beneficial use is: 'Beneficial use is the basis, the measure, and the limit of all rights to the use of water in this state ... consistent with the interest of the public in the best utilization of water supplies' (Beck, 1991). The tenets of the doctrine of effective and beneficial use are: (a) water is not to be obtained for speculation or let run to waste (reality of use); (b) the end use must be a generally recognized and socially acceptable use; (c) water is not to be misused (reasonable efficiency); and (d) the use must be reasonable as compared against other uses.

A common idea was that the quantity of water was to be no more than needed, the concern being with the possibility of 'vesting an absolute monopoly on a single individual' (Beck, 1991, pp. 107–8). This anti-monopoly/ anti-speculation concern where claimants do not have a specific use in mind continues today. For a long time it was difficult to assess what happens in practice when water legislation does not have a requirement of effective use, the reason being that national systems of water legislation did not normally grant exclusive non-riparian based water rights without adding the requirement of effective and beneficial use.

The Chilean experience on the issuance of non-conditioned water rights is an apparent validation of the concerns behind the requirement of effective and beneficial use. A study on the impact of the legal system for water allocation in Chile has found that it is common for formerly state-owned monopolies that benefited from exclusive rights to carry these rights when privatized, creating legal barriers to entry that maintain the monopolistic characteristics of the sector.

In Chile, the regulatory framework for electricity is based on the theoretical existence of competition in the generation of electricity. However, competition does not really exist and the water rights of the main hydroelectric projects belong mainly to a single corporation. The implication of this is that this generator has an incentive to appraise potential capacity investments, based on the effects that they will have on the profitability of its intramarginal capacity. It can achieve monopoly results over time by postponing investments. New entrepreneurs will be unable to enter the generation market because they do not have the water rights to undertake more efficient projects. Water rights should have been returned to the state prior to privatization, which in turn could have granted them subject to the conditionality of their timely development through new projects by existing producers or new comers (Bitran and Saez, 1993). Thus, the actual operation of the Chilean system appears to confirm the rationale behind the requirement of effective and beneficial use.

Empirical evidence on the actual working of water markets in Chile shows that, with a few local exceptions, market transactions of water rights in Chile have been limited (Bauer, 1998). Up to 1996, no more than 5 per cent of the water rights in highly utilized rivers had been transferred. And of this, 90 per cent of the transactions correspond to non-utilized rights, rather than to reallocations from one productive use to another (Pena, 1996).

The functioning of a strong judiciary is critical to the privatization process. In general, Latin American courts are traditional, inflexible and open to political and economic pressures. The Chilean Constitutional Court has recognized the right of the government to regulate the conditionalities of water rights (Rol 60/1997). In addition, the Anti-Monopoly Commission has recommended that no further water rights be granted until provisions requiring effective use of water are included into the water law (CPC 992/636; CR. 480/97).

In the Chilean case many mistakes were made with openness, transparency, participation, and ecosystems in the rush to get water markets set up. Once established, it is unlikely that performance can be improved over time by actually regulating the water rights and monopolies created, unlike the American experience that allows for adjustments when economic and social conditions change, because it does not aspire to build institutions that cover all possible eventualities (Rogers, 2002).

Unfortunately, the present situation of Latin America does not seem to endorse this kind of reasonable, pragmatic and flexible approach. To the contrary, the present status quo of water law, public utilities legislation, and agreements for the protection of investment emphasize unilateral security and contractual and legal strictures. Some argue that confiscation occurs when someone has to accept a return lower than expected, even if there still is a profit. Thus, it is not strange to see that water utility owners are guaranteed returns, rates of exchange and interest. This despite the fact that distinguished World Bank scholars have pointed out that these guarantees can wipe out the benefits of privatization by dampening incentives to select and manage programs and projects efficiently (Klein et al., 1998). In addition, these guarantees do in fact impose serious contingent liabilities on national budgets.

There are structural reasons for the deficiencies of regulatory systems in Latin America: a prejudiced view of government and a reliance on low information need has resulted in the design of weak information and follow-up systems, with reduced legal capabilities to gain information that would allow oversight and control of both operational and investment costs and the monitoring of transfer prices. Few systems have good regulatory accounting. The Chilean water utility regulator has realized this constraint, and is now requiring that companies provide more information on costs, expenses and income (*El Mercurio*, Santiago de Chile, Economía y Negocios, 13 April, 2002 p. B3). Companies are, predictably, challenging the need, usefulness and benefits of the request.

In the case of Argentina, it has been said that privatizations were sweet deals, with public utilities becoming private monopolies with rates in long-term contracts being updated according to American inflation, even if prices in Argentina were falling (Alcazar et al., 2000). It is claimed that the Buenos Aires Water Concession assumed that it was possible to approach water services according to paradigms developed for, and the technical characteristics of, more dynamic and innovative services, such as telecommunications and electricity. Tariffs were supposed to be fixed in real terms, for ten years, but during the first seven years of the concession there was a 45 per cent increase; coverage reached only 40 per cent of the population for water and 20 per cent for sewerage; and investment levels were lower than expected, while the reductions in investment were not reflected in lower tariff levels. According to Loftus and MacDonald, Aguas Argentinas has been making record profits with this concession, twice the international average and three times the UK water companies average (Rogers, 2002). The regulation model has been frail, inefficient and weak. The capture of the regulator and/or the government has been mentioned as one of the main reasons for the governance problems of the concession.

The River Rhine and Lake Constance: Enhanced Foreign Relations through Water Agreements

The evolution of governance in two important water-related regions in Europe, the River Rhine and Lake Constance, appears to exhibit a concern for international comity ahead of narrow water objectives (Blatter, 2003). The evolution of institutions has followed the diffusion of ideas rather than political power or market advantage. New challenges have been met by patching up existing institutions rather than switching over to new institutional forms.

The River Rhine is fed by Lake Constance, which is the second largest lake in central Europe. It is the only region in Europe in which the borders with Switzerland were never formally delineated. The lake serves as storage for potable water and a focus of tourism and recreation, especially boating. Institutionally, the lake is crosshatched with commissions and other agency structures, each attending to different subject matters and espousing various values. The Central Commission for Navigation on the Rhine River has existed continuously since 1815 despite wars and other dislocations. After World War II, France and Germany signed a treaty establishing the Commission for the Development of the Upper Rhine. Then, later, environmental concerns prompted the establishment of the International Commission for the Protection of the Rhine River Against Pollution and the International Commission for the Protection of Lake Constance. More recently, the movement toward European integration has spawned other organizations in the border region, including the German–French–Swiss Intergovernmental Commission for Border Issues. Further, while the form of institutions remained relatively unchanged, the operations and decision rules changed radically. Not only have local-level agencies come to have larger roles than national bureaucracies, but public/private interactions involving businesses and non-profit organizations have become much more common. Generally vertical relationships have been replaced by horizontal ones.

The changing logic for reaching agreement in the River Rhine and Lake Constance regions emphasizes a move away from universal principles like international law, water rights, or scientific and technical imperatives toward common identities and the building of shared beliefs. Because of the symbolic nature of water, water can be the basis of public engagement and participation in civic affairs and membership in transboundary groups that bridge national differences. Even the existence of many institutions and networks with overlapping and sometimes conflicting jurisdictions is not necessarily bad. The competition among institutions for membership and affiliates may lead to promising institutional innovations.

Examples of the Evolution of Urban Water Institutions in the Third World: Nepal, India and Thailand

The three case studies of urban water markets in Asia presented below highlight the growing role private sources of supply play in meeting the everyday needs of people in urban areas. Public water supplies, where available, remain highly subsidized but the poorest classes frequently have no service and end up paying very high prices to local water vendors while high-income customers are subsidized. Water quality is an increasing concern, especially for 'high-end' customers while little concern is shown for the environment.

Kathmandu
Nepal's largest city and capital is facing a stage of rapid development and expansion. These conditions started with the opening of the country in 1951 and the subsequent flood of modern development and population growth. Water supply in Katmandu was historically delivered through traditional systems, such as the widespread network of stone waterspouts or *dhunge dharas*. These stone spouts were part of a system for water supply that was supplied by *rajkhulos*, networks of drinking and irrigation water supply channels. The rapid development has led to the failure of a number of the *dhunge dharas*. Construction has cut off water supplies in places and the growing population generates more waste, contributing to the likelihood of the water's contamination. Although this traditional system is increasingly overwhelmed and polluted, it still supplies a significant portion of Katmandu's population

with water for domestic uses. In addition to stone waterspouts, many users also rely on local wells that have been dug or drilled into the upper aquifer underlying the city. These are generally viewed as polluted and the water is used for bathing, washing and other non-drinking uses.

Modern pumped water supply systems were introduced on a minor scale 100 years ago to provide water to royal and other high-status residences. This nucleus was subsequently expanded into a general municipal supply system, currently operated by the Nepal Water Supply Corporation (NWSC), which receives water from rivers flowing into the valley and a network of wells tapping lower confined or semi-confined aquifers beneath the urban area. The ability of this system to meet demands is limited. Overall, water supply from the municipal system is characterized by growing uncertainty and variation in the amounts delivered. During the dry season some households receive 0.5 to two hours of water a day, while others get water once a week or not at all. To compensate for shortages and losses, the government is investing in a major scheme, the Melamchi Project, to divert water from a stream outside the valley and deliver it to Katmandu through a 28-km-long tunnel. There have also been a variety of initiatives to reduce losses, with little effect to date. Regardless of these long-term plans, most current residents in the Kathmandu urban area experience significant shortages and disruptions in the supply they receive from the modern system.

As a result, the poor continue to depend on stone water spout systems and local wells tapping the increasingly polluted upper aquifer, while those who can afford to purchase water do so from what they hope are higher-quality vendor sources. Katmandu's private water market functions through tanker trucks that deliver water from a limited number of wells within the urban periphery to end-users. It feeds directly into the gap left by the municipal pipe system by delivering supplies reliably to private residences, hotels and other businesses in the valley. The tanker-based water market is, unlike in many other urban areas within South Asia, a relative luxury serving primarily the upper-middle and wealthy classes. The market functions outside the jurisdiction of the government and has no price or water quality regulations. As a result, the private market lacks official legitimacy and accountability, forcing individual companies to create their own standards to ensure the trust of their customers. Needless to say, there is a wide range of standards within the market and little means of verification.

Approximately 80 small tanker companies serve residents and commercial establishments dependent on the tanker water market and it meets as much as 18 per cent of total demand during the dry season. The net result is a mosaic in which the lower economic strata of society largely pay for water through women's labor, time, and the health consequences associated with pollution of wells and *dhunge dhara*. Upper levels of society pay through the direct cost of tanker water and also through potential health consequences associated with using water from unknown sources. Only a few sections of the city are able to rely on NWSC water for all their needs. These, generally wealthy, sections pay the lowest cost in both monetary and non-monetary terms.

Ahmedabad, Gujarat State, India

In Ahmedabad, physical access to water is not a major problem. The Ahmedabad Municipal Corporation delivers water twice a day for several hours. In dry seasons, deliveries to tap stands and households can occasionally be insufficient to meet basic domestic needs, but in most cases the volume of water available is sufficient.

Although the volume of supply available from municipal sources is, in most cases, sufficient and the price charged to consumers is highly subsidized, quality is a major concern. Ahmedabad residents have long been aware of the high total dissolved solids level in supplies available from groundwater sources under the city. Recent attention to excessive fluoride and the health problems associated with it has also increased concern over quality. The combination of short supply for large-volume users and low quality has driven the formation of a two-tier water market. Private companies with purification facilities sell partially demineralized water in pouches and bottles for users whose primary concern is quality, while private tanker companies deliver larger volumes.

Most private water vendors own their own source of supply, typically a borewell of 500–650 ft depth. They also typically operate their own delivery tankers. In some cases, however, well owners also contract with private tanker owners to deliver water to consumers for them. This is common in high demand periods, such as the marriage season, festivals and during droughts. Since these times of high demand are intermittent, water vendors avoid major capital costs by relying on intermediate-term delivery contracts or short-term tanker hires for these periods. Most suppliers have a set of fixed customers that include households, commercial establishments (hotels, restaurants) and, in some cases, industries. The market structure is significantly different from that in Katmandu in several aspects: (1) much of it is driven by quality rather than shortages; (2) there appear to be far fewer water vendors; and (3) low-end users don't face the large implicit costs that stem from having to wait hours at tap-stands to receive supplies.

Chennai, Thailand

Water supply for domestic use in Chennai urban area has been a source of concern for decades and, in recent years, the ability of the Metro Water Board to meet demand has fallen far short of available supply. Water deliveries are approximately half the government norm for urban water supply requirements in the Chennai urban area and only a small fraction of the demand that would probably be felt if supply were unrestricted and delivered at the highly subsidized rates found in other urban centers. Demand is also restricted because in water-short years piped water supply does not reach significant portions of the city on a regular basis. This situation has created the conditions for a flourishing and extensive water market in the Chennai urban area.

The tanker and private company market is highly fragmented. Numerous small companies run one or two tankers. They bring water either from their own wells or purchase it from farmers and other well owners. Many small purification companies are also present, each with their own facilities and each operating independent of any external check on the quality of the water they supply. Prices charged for water supply in the public and private sector vary greatly. Between the initial point of sale and the ultimate point of consumption, the price increases by many orders of magnitude. While this price increase reflects substantial service inputs (transport, purification, packaging, storage and cooling), the potential profits involved are very large.

Common themes

Several common themes run through these South Asian case studies, as described below.

Private sources have been growing and play a critical role in meeting many water demands. There is clearly a strong dynamic interaction between public and private sector supply systems. Private sector supply systems move into the gaps in service left by the public systems.

Water from public systems is generally highly subsidized and those subsidies are captured, in large part, by the wealthy and upper-middle classes. These groups are able to take whatever is available through the public system before purchasing water from outside sources. When they do need to purchase water, they buy in bulk and pay lower rates than those purchasing smaller quantities. Most of those purchasing both bulk water in tankers and purified drinking water in cans, carboys, bottles and packets are high-end consumers. Water from these sources, particularly the bottled drinking water, costs far more in monetary terms than that from public supplies. High-end consumers are increasingly willing to pay for reasons of convenience and quality, not, except in the most extreme instances, because water per se is unavailable at lower cost from other sources. In most cases, tanker markets and bottled water suppliers operate independent of any external check on the quality of the water they supply.

Although water markets predominantly serve high-end consumers, low-end consumers often pay a far higher real price for water, if the time and labor they must spend in obtaining it is taken into account. Furthermore, the quality of low-end supplies is often more open to question than that for high-end users. Only at the highest end of the market, that for bottles and packets of water, are the wealthy paying more on a regular basis than the poor. The water markets function in the absence of any formal rights or regulatory system designed to protect the resource base.

Rural-to-rural water transfers through market-like mechanisms tend to be common in South Asia. Whether or not these transfers are fair and equitable depends a good deal on the power structure, and there can be coercive transfers among rural residents, especially if jurisdictional boundaries are crossed. Some transfers can force marginal farmers deeper into dependency relationships. Rural-to-urban water transfers are becoming more frequent in South Asia, and promise to play a larger role in addressing projected urban shortages. There is no question that such transfers promote flexibility and probably efficiency in water supply, but equity may be a different matter. Whether water is delivered publicly or privately, higher-income urban residents seem to be advantaged because they have higher capacity to store water and because they are well served by subsidized public water rates.

Well-established property rights seem not be a prerequisite for water markets in South Asia where access to groundwater and energy to pump seem to be the only prerequisites. Other institutional arrangements generally thought to be prerequisites for well-functioning markets do turn out to be important in South Asia as well as Latin America. Moench (2002) and Bauer (1998) agree that courts overseeing the operation of private water deliveries and water markets must be strong, independent and willing to make substantive as well as procedural judgments.

Experience of poor people in South Asia is one of extreme scarcity of easily accessible, healthful water for domestic use. In Katmandu, it is not so much scarcity as it is the inconvenience and poor quality of water available to low-income families that is the most serious problem. Women wait for as long as seven hours in order to get water from public water spouts, which they must then carry to their homes. Markets have done little to ease the burden on the poor who pay a disproportionate share of their income for unreliable supplies of questionable quality. On the other end of the economic scale, middle and upper-class household that have storage facilities as a buffer against unreliable and intermittent public water supplies receive water at artificially low water rates, which discourage conservation practices.

3.   CONCLUSIONS AND RECOMMENDATIONS

Many countries around the globe are rethinking the roles of the public sector and private enterprise in the provision of public services. This is also taking place in the American West. As pointed out by James Wescoat (Chapter 1), the western United States was an early innovator in terms of devolution of water resource functions to lower levels of government and the contracting out of services to private enterprise. Today, however, other nations and regions are experimenting with new institutional forms and operating practices that sometimes outpace what is happening in the American West. This is seen in several substantive areas: for example, the privatization of municipal water and wastewater services, a return to the river basin as the basic unit of management, large-scale reallocation of water to environmental purposes, and the use of water markets for the distribution of raw water. These steps are motivated by the shortage of funds for badly needed construction and maintenance of infrastructure and a need for greater expertise in meeting regulations of increasing complexity. Often there is the sense that bureaucracies are out of touch with public values and/or are the source of unreasonably burdensome regulations.

While a major objective of this chapter is to identify policy changes that will benefit the western United States, some of the lessons learned have global applicability. Thus some of the lessons noted below may already be in place in the western United States but are sufficiently valuable to be cited for the broader benefit. After this review of general lessons, our attention returns to lessons specifically relevant to the western United States – again with the acknowledgement that most of these lessons may also have broader relevance. We conclude by addressing how change and innovation in government and private sector responsibilities might be achieved.

General Lessons

An overarching lesson emerges from experience around the world: *water serves multiple values* some of which are poorly reflected in standard market transactions. Experience has shown that equity must be built into program and project designs rather than being left to other programs or social safety nets that frequently do not exist. Cultural values should over-ride economics in some settings. These value differences imply the need for participatory, transparent decision processes in which monetized values of the usual benefit–cost type must be weighed against all the other values. Benefit-cost analysis should be a part (but not all) of the decision process.

Any move towards privatization should also take place within a transparent, accountable framework that has the capacity to monitor system performance and to protect public values. Markets in general must meet several key requirements if they are to operate in a 'socially efficient' manner: full information; multiple buyers and sellers; well-defined property rights; and equality of strength in bargaining. These conditions apply to water markets also, but are not likely to be met in many cases. Thus continuing social monitoring and oversight through governmental and political institutions must continue indefinitely.

Attempts to privatize ownership of the basic water resource through the establishment of tradable water rights have had mixed results. In Chile, river flows have been monopolized by large companies, which were allowed to register for indefinite-lived rights without the demonstration of a 'beneficial use' and in the absence of a strong judiciary that could uphold existing laws and hear appeals (Bauer, 1998; Solanes, 2002). In the case of urban water for Buenos Aires, the company chosen was not closely monitored, failed to provide the expansion of service promised, and had negotiated a system of rewards that continued independent of contract fulfillment (Laborde, 2002; Solanes, 2002). Thus, *the 'beneficial use doctrine' and a strong judiciary are necessary conditions for the effective privatization of the basic resource and urban services.*

That private provision of water services should and can follow the patterns of deregulation successfully followed in other public services such as electric power, gas, and telecommunications is in serious error. The development of new technologies (such as dish reception of satellite signals, remote monitoring and cell phone coverage) and new institutional arrangements (required sharing of existing telephone, gas and electric distribution systems, the elimination of interstate pipeline regulations) have overcome the 'natural monopoly' character of some of these services. However, *the provision of urban water services remains a 'natural monopoly' that must be monitored and regulated in the public interest.*

The river basin remains the natural unit for water development and management, yet few countries have historically protected the unity of river basin administration. Unfortunately, the geographical locations of political and community interests often do not coincide with those of river basins, thus complicating governance. Even in the face of interjurisdictional compacts, this disjunction of physical and managerial boundaries leads to serious 'jurisdictional externalities', that is negative impacts on other parties that are not taken into account by the water management agencies. Both Australia and Brazil are trying to return to river basin administration. In addition, in a variety of settings including the American West, watershed partnerships are showing real vitality.

The creativity of people in resolving issues and avoiding open conflict should not be underestimated. Wolf (1998) has convincingly shown that *finding accommodation in the allocation of water across borders and within local jurisdictions can create a new sense of community.* In the Middle East there has been more cooperation than conflict over water.

Water and land use must be dealt with together in planning, permitting and regulatory processes. Water use and water-borne waste generation are both tied to land use. The largest remaining sources of organic water pollution in many countries are the agricultural and livestock industries, while mining (usually old, abandoned mines) is the major source of heavy metal pollution. Regional development is frequently allowed to proceed without guarantees of long-term water supply. Land use planning is seldom used in the western United States and is almost never coordinated with water planning.

The fundamental needs of the poorest of the poor must be kept in mind in designing governmental and market institutions. Whether the distribution of municipal water is through government or private utilities, the record of meeting the needs of the very poor has generally been unsatisfactory (White et al., 1972). Much more attention needs to be paid to the design of simple delivery systems and methods of payment that will facilitate rapid expansion of potable water services sufficient for healthful living.

Conclusions and Recommendations Specific to the Western United States

National agricultural policies underlie many of the supply–demand and drought stress problems faced in the western United States. *The continued subsidization of major crops through price support programs and underpriced irrigation water stands in the way of increased efficiency in national water use.* Western sugar beet production, irrigated rice and short staple cotton have little value to the nation as a whole since they can be produced more cheaply in other regions or replaced by imports. The volume of water consumed by these crops alone is several times the amount consumed by cities in the West. A gradual shift in regional cropping patterns and the elimination of non-competitive crops with transitional payments to facilitate human adjustment is a long-term goal, but must not be lost sight of as short-term crises are addressed.

The motivation for the individual farmer, ditch company or irrigation district to adopt conservation measures is often weak or totally absent when farmers pay low prices for irrigation water. Even if long-term contracts cannot be changed, the presence of active water markets will signal the real opportunity cost to the farmer. *Gradually raising contract prices for project water and simplifying water marketing procedures should be pursued.* Innovative water transfer programs allow urban areas to maintain the reliability of their supplies, while leaving the long-term ownership of the supply in agriculture. The 'drought-year lease-out' is one such mechanism. City water managers have expressed interest in these arrangements in lieu of the purchase of additional water rights, provided the contract has an indefinite terminal date, while the facilitation of short-term transfers within the agricultural sector through 'water banks' can increase farm incomes and help hold water 'down on the farm'.

A major opportunity for stretching water supplies, increasing reliability of supplies, and enhancing the riparian environment is through *greatly expanded programs of conjunctive management of groundwater and surface water.* Using renewable groundwater has obvious advantages of natural distribution to the point of use, little or no evaporative losses during storage and availability during droughts. Interconnections with surface flows must be recognized and compensated. This introduces problems during extended droughts when compensatory surface water may not be available (MacDonnell, 1988).

The coordination of water quantity allocation, land use and water quality policy is an old theme that has been largely ignored in the western United States. Every use of water changes its quality, yet in some western states water rights are administered solely in terms of the quantity or (flow rate) of water to be diverted. The quality of the return flow is frequently ignored even though some areas produce return flows laden with salts or suspended matter.

How Do We Effect these Institutional Changes?

Adaptive management requires adaptive institutions. The recommendations noted above may be difficult to carry out and difficult to adapt to particular situations. The historical record of change in water policy, like many other policy areas, has evidenced long periods of stability punctuated by cycles of innovation and change. The American West was a pioneer in addressing water problems through large-scale physical structures starting in the early 1900s and dominating the next half century. In recent years far more attention has been directed towards water conservation, reallocation, and transfer. Water policy makers tend to be conservative in outlook and would prefer a low profile and an absence of controversy (Rayner et al., 2002). While the western United States has incrementally and somewhat cautiously moved in the direction of new water allocation mechanisms like water markets, the pace of change has been halting and the pattern mixed.

Change is most likely to take place when there is visible evidence that present governance is inadequate for new conditions. Such evidence may come in many forms including a triggering event such as a severe, sustained drought, evidence of severe environmental damage, domestic water quality problems, an unsettling court decision or dramatic changes in political control. As noted by Hayes (2002), it usually requires a crisis to get public backing for significant institutional change but individual leadership is always important, as illustrated by the involvement of Secretary of Interior Bruce Babbitt and his office in the development of the CALFED project. Change becomes institutionalized and actually implemented when the most affected and influential interests see themselves as benefited.

Many opportunities will arise to experiment with water allocation mechanisms, given the emerging water problems and heightened public concern about health and the environment. We anticipate that the next two decades will evidence a burst of institutional innovation informed by the global experiences cited above.

LITERATURE CITED

Agencia Nacional de Aguas (ANA) (Brazil) (2002), *The Evolution of Water Resources Management in Brazil*, Brazilia: ANA, March.

Alcazar, Lorena, Manual A. Abdala and Mary M. Shirley (2000), 'The Buenos Aires concession', The World Bank Development Research Group policy research working paper 2311, April.

American Water Works Association (AWWA) (1992), *Water Industry Data Base: Utility Profiles*, Denver: AWWA.

Anderson, Terry L. and Donald R. Leal (2001), *Free Market Environmentalism*, New York: Palgrave.

Barlowe, Raleigh (1983), 'Changing land use and policies: the Lake States', in S.L. Flader (ed.), *The Great Lakes Forest: an Environmental and Social History*, Minneapolis, MN: University of Minnesota Press.

Bauer, Carl J. (1998), *Against the Current: Privatization, Water Markets, and the State in Chile*, Boston: Kluwer Academic Publishers.

Baumgartner, Frank (2002), 'Social movements and the rise of new issues', paper presented at the University of California – Irvine conference on Social Movements, Public Policy, and Democracy, 11–13 January.

Beck, Robert E. (ed.) (1991), *Water and Water Rights*, vol 2, Charlottesville, VA: The Michie Company.

Bitran, Eduardo and Raul Saez (1993), 'Privatization and regulation in Chile', Brookings Institution conference on the Chilean Economy, Washington, DC: Brookings.

Blackmore, D.J. and C.J. Ballard (2002), 'Interstate water trading: current opportunities and barriers: the Murray–Darling Basin experience', paper prepared for the Natural Resources Law Center, University of Colorado School of Law, Boulder, CO, June.

Blatter, Joachim (2003), 'Beyond hierarchies and networks: institutional logics and changes in transboundary political spaces during the 20th century', *Governance*, **16**(4), 503–26.

Blomquist, William A. (1992), *Dividing the Waters: Governing Groundwater in Southern California*, San Francisco: ICS Press.

Brown, F. Lee and Helen M. Ingram (1987), *Water and Poverty in the Southwest*, Tucson, AZ: University of Arizona Press.

Cortner, Hanna J. and Margaret Moote (1999), *The Politics of Ecosystem Management*, Washington, DC: Island Press.

Costanza, Robert (ed.) (1991), *Ecological Economics: the Science and Management of Sustainability*, New York: Columbia University Press.

Crawford, Stanley (1988), *Mayordomo: Chronicle of an Acequia in Northern New Mexico*, Albuquerque, NM: University of New Mexico Press.

Doremus, Holly (2001), 'Adaptive management, the Endangered Species Act and the institutional challenges of "New Age" environmental protection', *Washburn Law Journal*, **41**(1), 50.

Doyle, Mary and Donald Jodrey (2002), 'Everglades restoration: forging new law in allocating water for the environment', *Environmental Lawyer*, **8** (February), 255–302.

Gleick, Peter H. (2001), *The World's Water 2000–2001: the Biennial Report on Freshwater Resources*, Washington, DC: Island Press, chapter 4.

Gopalakrishnan, Chennat (2002), 'Private gain at public loss: the political economy of water allocation and use in Hawaii', proceedings of the International Water History Association, accessed at (www.iwha.net/abstracts).

Hays, Samuel P. (1959), *Conservation and the Gospel of Efficiency*, Cambridge, MA: Harvard University Press.

Hayes, David J. (2002), 'Federal–state decision-making on water: applying lessons learned', paper prepared for the Natural Resources Law Center, University of Colorado School of Law, Boulder, CO, June.

Hobbs, Gregory J., Jr. (2002), 'Priority: the most misunderstood stick in the bundle', *Environmental Law*, **32**, 32–58.

Howe, Charles W. (1988), 'Water as an economic commodity', in David H. Getches (ed.), *Water and the American West*, Boulder, CO: Natural Resources Law Center, University of Colorado School of Law, chapter 4.

Howe, Charles W. (1997), 'Dimensions of sustainability: geographical, temporal, institutional, and psychological', *Land Economics*, **73**(4), 597–607.

Howe, Charles. W. (2000), 'Protecting public values in a water market setting: improving water markets to increase economic efficiency and equity', *University of Denver Water Law Review*, **3**(2), 357–72.

Howe, Charles W. and Mark Griffin Smith (1994), 'The value of water supply reliability in urban water systems', *Journal of Environmental Economics and Mangagement*, **26**, 19–30.

Howe, Charles W., Jeffrey K. Lazo and Kenneth R. Weber (1990), 'The economic impacts of agriculture-to-urban water transfers on the area of origin: a case study of the Arkansas Valley in Colorado', *American Journal of Agricultural Economics*, **72**(5), 2300–4.

Howe, Charles W., Dennis Schurmeier and W. Douglass Shaw, Jr (1986a), 'Innovations in water management: lessons from the Colorado-Big Thompson Project and Northern Colorado Water Conservancy District', in Kenneth D. Frederick (ed.), *Scarce Water and Institutional Change*, Washington, DC: Resources for the Future, Inc.

Howe, Charles W., Dennis Schurmeier and W. Douglass Shaw, Jr. (1986b), 'Innovative approaches to water allocation: the potential for water markets', *Water Resources Research*, **22**(4), 493–545.

Howe, Charles W., J.L. Carroll, A.P. Hunter, W.J. Leininget et al. (1969), *Inland Waterway Transportation; Studies in Public and Private Management and Investment Decisions*, Washington, DC: Resources for the Future.

Ingram, Helen (1987), 'The community value of water: implications for the rural poor of the Southwest', *Journal of the Southwest*, **29**(2), 179–202.

Ingram, Helen (1990), *Water Politics: Continuity and Change*, Albuquerque, NM: University of New Mexico Press.

Kaiser, R.A. and L.M. Phillips (1998), 'Dividing the waters: water marketing as a conflict resolution strategy in the Edwards aquifer system', *Natural Resources Journal*, **38**(3), 411–44.

Keplinger, Keith O. and Bruce A. McCarl, (1996), 'Conjunctive management implications for the Edwards aquifer: an integrated aquifer/river basin economic optimization approach', in *Proceedings: Integrated Management of Surface and Groundwater*, Universities' Council on Water Resources, Carbondale, IL: Southern Illinois University.

Klein, Michael, Timothy Irwin and Matee Thobani (1998), *Dealing with Risk in Private Infrastructure*, Washington, DC: The World Bank.

Laborde, Lillian del Castillo de (2003) 'Institutional framework for water tariffs in the Buenos Aires concession', in Cecilia Tortajada and Asit Biswas (eds), *Water Pricing and Public–Private Partnership in the Americas*, Washington, DC: Inter-American Development Bank, pp. 97–117.

Lochhead, James S. (2001), 'An upper basin perspective on California's claims to water from the Colorado River, part 1: the law of the river', *University of Denver Law Review*, **4** (Spring), 290–330.

Maass, Arthur (1957), *Muddy Waters: The Army Engineers and the Nation's Rivers*, Cambridge, MA: Harvard University Press.

Maass, Arthur and Raymond J. Anderson (1978), *...And the Desert Shall Rejoice*, Cambridge, MA: MIT Press.

MacDonnell, Lawrence J. (1988), 'Colorado's law of underground water: a look at the South Platte and beyond', *University of Colorado Law Review*, **59**(3), 579–625.

MacDonnell, Lawrence J. (1989), 'Changing uses of water in Colorado: law and policy', *Arizona Law Review* (Water Transfer Symposium), **31**(4), 783–816.

MacDonnell, Lawrence J., Charles W. Howe, Kathleen A. Miller, Teresa A. Rice and Sarah F. Bates (1994), *Water Banks in the West*, Boulder, CO: Natural Resources Law Center, University of Colorado School of Law.

MacDonnell, Lawrence J. and Teresa Rice (1994), 'Moving agricultural water to cities: the search for smarter approaches', *Hastings West-Northwest Journal of Environmental Law and Policy*, **2**(1), 27–55.

Major, David C. and Roberto L. Lenton (1979), *Applied Water Resource Systems Planning*, Englewood Cliffs, NJ: Prentice-Hall.

Mapp, Jr, H. (1972), 'An economic analysis of water use regulation in the Central Ogallala Formation', PhD dissertation presented at Oklahoma State University, Stillwater, OK, May.

Meyer, Michael C. (1996), *Water in the Hispanic Southwest: a Social and Legal History 1550–1850*, Tucson, AZ: University of Arizona Press.

Michelsen, Ari M. (1994), 'Administrative, institutional and structural characteristics of an active water market', *Water Resources Bulletin*, **30**, 971.

Moench, Marcus (2002), 'Water markets, commodity chains and the value of water', paper prepared for the Natural Resources Law Center, University of Colorado School of Law, Boulder, CO, June.

Moncur, James E.T. (1989), 'Economic efficiency and institutional change in water allocation: the 1987 Hawaii Water Code', from the proceedings of the 25th Annual Conference, American Water Resources Association, Tampa, 17–22 September.

Musgrave, Richard A. (1959), *The Theory of Public Finance: A Study in Public Economy*, Ann Arbor, MI: University of Michigan Press.

National Research Council (2002), *Privatization of Water Services in the United States: an Assessment of Issues and Experience*, Washington, DC: National Academy Press.

Oggins, Cy R. and Helen Ingram (1989), *The Community Consequences of Rural to Urban Water Transfers*, Tucson, AZ: The Udall Center for Studies in Public Policy.

Pena, Humberto (National Water Director of Chile) (1996), 'Foro del Sector Sanamiento sobre el Proyecto de Ley General de Aguas', texto de expositiones, Lima, Peru, 8/9 January.

Powell Consortium (1995), *Severe Sustained Drought: Managing the Colorado River System in Times of Water Shortage*, Tucson, AZ: Arizona Water Resources Research Center.

Rayner, Steve, Denise Lach, Helen Ingram and Mark Houck (2002), 'Weather forecasts are for whimps: why water resource managers don't use climate forecasts', final project report to the National Atmospheric and Oceanographic Program, Washington, DC.

Renwick, Mary and Sandra Archibald (1997), 'Demand side management policies for residential water use: who bears the conservation burden?', paper given at the annual meeting of the Western Regional Science Association, Big Island, Hawaii, February.

Rice, Teresa A. and Lawrence J. MacDonnell (1993), *Agricultural to Urban Water Transfers in Colorado: an Assessment of the Issues and Options*, Boulder, CO: Natural Resources Law Center, University of Colorado School of Law.

Rogers, Peter (2002), 'Water governance', (Fortaleza, Brazil, draft), paper prepared for the Inter-American Development Bank, 4 February.

Solanes, Miguel (2002), 'The privatization of water in Latin America: water rights, the beneficial use doctrine and regulatory failure', paper prepared for the Natural Resources Law Center, University of Colorado School of Law, Boulder, CO, June.

State of California (1991), 'Conceptual approach for reaching basin states' agreement on interim operation of Colorado River System reservoirs, California's use of Colorado River water above its basic apportionment, and implementation of an interstate water bank', paper prepared for the Colorado River Basin States' meeting by the Colorado River Board of California, 28 August.

Tyler, Daniel (1992), *The Last Water Hole in the West: the Colorado-Big Thompson Project and the Northern Colorado Water Conservancy District*, Niwot, CO: University of Colorado Press.

United States Geological Survey (USGS) (1998), 'Estimated water use in the United States in 1995', USGS Circular 1200, Washington, DC: US Government Printing Office.

Vaux, H.J. and Richard E. Howitt (1984), 'Managing water scarcity: an evaluation of interregional transfers', *Water Resources Research*, **20**, 785.

Wahl, Richard W. (1989), *Markets for Federal Water: Subsidies, Property Rights and the Bureau of Reclamation*, Washington, DC: Resources for the Future, Inc.

Weber, Kenneth R. (1989a), 'Social and economic consequences of water sales in Crowley County, Colorado: a background paper', unpublished paper, Environment and Behavior Program, Institute of Behavioral Science, 20 July.

Weber, Kenneth R. (1989b), 'What becomes of the farmers who sell their irrigation water? the case of water sales in Crowley, County, Colorado', unpublished paper, Environment and Behavior Program, Institute of Behavioral Science, 16 November.

Weber, Kenneth R. (1990), 'Effects of water transfers on rural areas: a response to Shupe, Weatherford and Checchio', *Natural Resources Journal*, **30** (Winter), 13–15.

Wescoat, James L. Jr (1989), 'Picturing an early mughal garden', *Asian Art*, Fall, 59–79.

68   *In search of sustainable water management*

White, Gilbert F., David J. Bradley and Anne U. White (1972), *Drawers of Water: Domestic Water Use in East Africa*, Chicago: University of Chicago Press.
Wolf, A. (1998), 'Conflict and cooperation along international waterways', *Water Policy*, **1**(2), 251–65.
World Commission on Environment and Development (Brundtland Commission) (1987), *Our Common Future*, Oxford and New York: Oxford University Press.
Young, Mike, Darla Hatton MacDonald, Randy Stringer and Henning Bjornlund (2000), 'Interstate water trading: a two year review', draft final report to the Commonwealth Scientific and Industrial Research Organisation Land and Water, December.

3. Integrating environmental and other public values in water allocation and management decisions

David H. Getches and Sarah B. Van de Wetering

With assistance from David Farrier, University of Wollongong; Robyn Tanya Stein, Bowman Gilfillan Inc.; and Wang Xi and colleagues, Wuhan University

1. REVIEW OF ISSUES IN THE AMERICAN WEST

The Problem: Protecting Public Values in Systems of Private Rights

In the American West, as in most societies of the world, water is a public resource. Private users of water must obey rules designed to protect and enhance broad public values. The rules are rarely rigorous and what is included within the definition of 'public values' varies. Most people understand public values to include the important services that water provides that are difficult to quantify in monetary terms. A few examples illustrate the diverse services provided by water in the arid and semi-arid American West: preservation of biological diversity in healthy, functioning aquatic ecosystems; opportunities for aesthetic appreciation and spiritual renewal; recreational activities such as fishing, boating, swimming, bird watching, hiking, and scenic driving; cultural identity and historical activities related to streams and lakes; and concerns for future economic opportunities dependent on reliable water supplies. Some of these services support commercially valuable activities and industries; others lack tangible monetary value but promote the well-being of society.

Since the first non-Native people came to the United States, the task of water law has been to allocate water to individuals and enterprises for utilitarian purposes and thus further a broad public interest in economic expansion. Until it was so allocated, water remained a public resource. Private uses allowed by law, however, could come into conflict with public uses. For example, occasionally the public's right to use waterways for boating clashed with uses that obstructed or depleted the flow of streams, and water contamination impaired domestic water uses. Only in recent years has the law recognized that the rights of the public can be strong enough to trump individual economic uses.

Two phenomena have combined to create a greater consciousness of the public's interest: (1) scarcity of water sources and (2) an increased understanding of the interconnection of forms of life (including human life) that are dependent on ecosystem health. Scarcity of water in the United States results in part from a growth in urban demand and in part from the demand for free-flowing streams to sustain ecosystems and for human recreational uses. Because most of the useful sites for major water projects have been developed or are off-limits for environmental reasons, the old response to water problems – developing new sources by constructing water projects – is no longer viable.

Most states' early water laws included nominal protections for the public interest, but historically these public rights had little significant impact on the fulfillment of utilitarian purposes. Recently, however, state and federal laws have given more substantial, if uneven, protection to the public interest. Today, there are many legal mechanisms for protecting the public interest. Some are imposed at the point when waters are appropriated or when an existing use is changed. Some apply when a dam or other major structure is built. Federal environmental laws have proven the strongest mechanisms for asserting and protecting diverse public interests.

In this section we examine the roots of competition between public and private rights. The following section examines the means that have been adopted by state and federal legislatures and, in some cases, by courts to ensure that public interests are protected, with a particular emphasis on environmental protection.

Legal protections for private uses of water

Water law developed differently in the eastern and western parts of the United States. Despite their considerable differences, both water rights systems aimed to provide basic human necessities and favored the productive use of water by permitting citizens to establish private rights and to protect their uses under the law. Neither system imposed any monetary charge for the right to divert water from a stream and put it to use for private gain.

In the relatively humid eastern United States there was little need to move much water out of watercourses. Typical early uses of water in the East included mills powered by waterwheels and modest domestic uses of stream water. Consequently, water law in that region involved equitable sharing among all riparian users. Landowners along streams and adjacent to lakes held 'riparian rights' to the waters for the benefit of their lands. Although 'pure' statements of the law said that the landowner's right was to the full flow of the stream, undiminished in quantity and quality, the law virtually always allowed water to be diverted from the stream by other riparian land-owners for uses on their lands. The one enduring rule of the riparian doctrine is that landowners along a watercourse must share the water equitably, as expressed in the 'reasonable use' doctrine.

In the more arid western United States economic activities such as mining and agriculture required that water be moved far from riverbeds. Rights to use water followed a rule of capture already applied in California mining camps that became known as 'prior appropriation' – the idea that the first person to put water to use should have a right to continue that use regardless of subsequent diversions by others, upstream or downstream. The prior appropriation doctrine explicitly allowed an appropriator to use water far from the stream of origin – even in another watershed – and did not contain any prohibition on diverting a stream's entire flow if that was necessary to satisfy a senior water right (Bates et al., 1993).

In those days the land was almost entirely federal property, 'public land', but the federal government did not dictate the manner of allocating water rights. Indeed, it allowed states to create their own systems of assigning rights to use water located on government land. In this way, the government encouraged investments in irrigation systems to serve lands granted to private parties by the government and investments for mines located on public lands. Like a subsidy, giving private parties free rights to use water created an incentive to economic activity consistent with the government's desire to promote western expansion and local economic development.

Given the settlement patterns in this region, other departures from eastern water law were necessary. For example, an appropriator did not need to be a landowner as most of the land in the West in the 19th century was the property of the federal government until it was patented (transferred) to homesteaders, miners, states, or railroads under various government programs. Rights could be lost, however, if the holder of the rights discontinued the beneficial use.

Today, rights in both eastern and western states are managed largely under systems in which administrative agencies issue permits and administer water use according to codified rules. Not every state has a permit system; Colorado, for instance, uses courts in place of administrative agencies to determine the extent of water rights. Although there are marked differences among state programs, rights to use water – even under administrative permit systems – are generally considered a form of property. The property in water use is not a possessory right like rights to real or personal property that allow the owner to exclude others' access or use. It is, instead, a 'usufructuary right', a right to use water in ways that are consistent with the public's interests.

The West's prior appropriation system did not restrict water rights to use on a specific plot of land or to a specific type of use. Instead, one user could transfer water rights to another and could use water wherever technology and economics would let it be moved. Thus, as the region's economies matured, water rights systems could adapt to satisfy growing and competing demands. Even in the East, where water rights were ostensibly attached to specific land, courts made exceptions to allow water use on non-riparian lands in order to satisfy economic and social demands.

An appropriator's capacity to transfer a water right – that is, to transfer legal priority to use a quantity of water for a beneficial purpose – is a fundamental element of the 'property' that the law recognizes in water use (see Howe and Ingram, Chapter 2 of this volume). The transfer of water rights is subject to the condition that a change of use should not damage the water rights of any other water user. This 'no injury' rule is the only universal restriction against water transfers. Initially some states also limited transfers with other restrictions, but those constraints have largely fallen with the need to move water from agricultural to urban uses. Meanwhile, legislatures have imposed other restrictions that were considered necessary to protect the public interest.

Public interests and values

When water is allocated, developed, or transferred, it can affect interests or values of the public, potentially depriving existing users of quantities of water, changing the flow of streams, or affecting water quality. The effects include:

- environmental impacts, including reduced streamflows, loss of wetlands, damaged ecological systems including fish, wildlife, and riparian vegetation, and diminished water quality;
- harm to other public values that are difficult to quantify, such as aesthetics and loss of recreational opportunities;
- economic and social effects, including the loss of income, employment, and business opportunities when a community loses its nearby source of water.

Even pumping groundwater can have adverse effects. Extraction of groundwater can alter surface flow with potentially adverse effects on vegetation and riparian habitat (Glennon, 2002). The impacts are most frequently realized in the American West today as agricultural water rights are converted to urban uses. Drying up formerly irrigated lands can lead to soil erosion and blowing dust and the invasion of noxious weeds.

Historically, the emphasis on encouraging private investment and settlement in the West meant that these impacts received little consideration and had no means of legal protection. The 'pure' prior appropriation doctrine required that water be physically removed from a stream and put to a recognized beneficial use in order for the user to claim a legally protected water right. Thus, those who enjoyed the instream benefits of water – for example, recreational boaters, anglers, and resort owners dependant on scenic vistas – could not prevent those coming later from diverting water from the stream. Similarly, courts and administrative agencies did not attempt to protect broader public interests in water when early water users established the most senior water rights. This is slowly changing. The next section describes how western water laws address public interest values.

Public Interest Protection

Generally, three types of activity affect the public's interest in streams: (1) diversions that deplete streamflows; (2) structures such as dams that obstruct streams and change flow patterns and temperatures; and (3) discharges that pollute waterways. The first type of activity falls within the traditional ambit of state law. The other two are largely regulated by federal law. The United States historically has deferred to the states and allowed them to allocate quantities of water according to their own laws. More recently, the federal government has established environmental protection programs that can interfere with water use under state law. This has raised federalism issues. As a legal matter, the federal government has ample power to preempt state laws, but there is political resistance to national legislation and administrative actions that conflict with state water laws. Although state laws recognize water as a public resource, states have done little to protect many public values.

This failure cannot be attributed to lack of legal authority. The earliest state legislation or constitutional provisions asserted that grants of private rights in water must be consistent with the public interest or public welfare. Other laws were more specific in preserving particular social or economic interests, especially agriculture. For instance, some laws said that agricultural uses should be protected and speculation should be prevented in water. Other state laws made water rights appurtenant to specific lands and flatly prohibited their transfer.

Typically, there is no forum for private individuals and groups to enforce legal provisions that appear to protect the public's interest in water. Public agencies usually receive comments from parties directly involved in decisions concerning the allocation, development, and transfer of water rights. Modern laws have begun to open decision-making processes to representatives of various interests who are affected by the allocation or use of water rights. Sometimes members of the public can comment, but unless they have water rights they are 'third parties' to the transaction without substantial protection for their interests.

Opportunities to protect the public interest arise in the water decision-making process when water is allocated to new uses, new projects are proposed, and when water uses are changed or transferred. The discussion below describes several methods to integrate public values and assesses their effectiveness. We conclude that there is considerable potential for the programs now in use in the United States but that they now provide incomplete protection for the interests of the public.

Public interest review

Most water rights systems attempt to limit or prevent adverse impacts on the public from water uses at the time new users obtain permits or when water rights are changed or transferred. This type of protection, however, is neither universal nor adequate in most states because processes are sometimes limited to water rights holders and exclude individuals and other entities that experience economic, environmental, and social impacts from water use and development. There is pressure, however, in most jurisdictions to include all such affected interests in the process of determining whether the public interest is served by a proposed water decision.

Although state laws often stipulate that water allocation must be consistent with the 'public interest' or 'public welfare', in practice, states rarely deny new uses or transfers in order to protect the public interest. Instead they impose additional conditions on the appropriation or transfer. Some, but not all, states apply the same public interest requirements to changes of use or transfers that they impose on new appropriations.

The Supreme Court of Utah upheld the application of the same criteria to changes in use that it applies to new appropriations.[1] In Nevada, a statute requires the state to reject an application for a water transfer that would result in damaging the public interest.[2] Wyoming, one of the few states with a special process to evaluate transfers, considers potential economic losses to the community relative to the benefits of the transfer and the availability of other sources of water.[3] California's State Water Resources Control Board reviews proposed transfers to determine if they would cause an unreasonable effect on the economy in the area of origin or on fish, wildlife, or other water uses.[4]

Almost every state in the West, except Colorado, uses some type of process to review the public interest in water decisions; all could improve the way in which they review the effects. The majority of the states lack clear standards to define the public interest that they are trying to protect. For example, New Mexico's statute simply directs the state engineer to determine whether a water use would be 'detrimental to the public welfare'. By contrast, Alaska's statute is more detailed, listing eight specific factors to be considered in a public interest review. In other states, vague statutes have been supplemented by judicial interpretations. For instance, state law requires Idaho's Department of Water Resources director to determine whether a proposed water use is in conflict with 'the local public interest', but the statute does not define this standard.[5] The Idaho Supreme Court has read the statute with reference to other laws of Idaho and of other states that define the public interest.[6] Following that ruling, the Director of Water Resources has convened hearings aimed at reaching decisions that ensure 'the greatest benefit possible to the public [from public waters] for the public'.[7] Affected citizens can present evidence about matters such as aesthetics, recreation, fish, and ecosystem functions that will be impacted by the proposed water decision. The agency considers not only benefits to the applicant but also economic effects, alternative uses, minimum streamflows, wastewater, and conservation.

Still, most state agencies do not fully consider many of the social, economic, and ecological interests affected by water allocation, transfer, and use. If the elements constituting the public interest were comprehensively articulated, government employees could use them as a guide for state policy in resolving conflicts among competing interests and to understand better the trade-offs inherent in any water decision. Comprehensive water planning – of which there is little in the American West – could help articulate both the elements of the public interest and state policies related to them.

Area of origin protection laws
Local governments, Indian tribes, and rural communities near water sources frequently suffer the greatest effects when water decisions benefit more populous areas or interests with greater political or economic power. For example, water removed from nearby water sources and used elsewhere may inhibit the future development of local communities. Water taken out of existing agriculture and transferred elsewhere may reduce agricultural employment in the area and impact agriculture-related businesses. Local municipalities required by law to maintain a certain water quality may find that reduced streamflow will increase their costs of treating sewage because it is more difficult to dissolve adequately the discharged waste. As tax bases decline and local businesses suffer there is a resulting decline in the ability of the local government to provide services to citizens. The area, in turn, becomes less attractive to new businesses. Social impacts of water allocation, development, and transfer therefore include changes in community structure, cohesiveness, and control of natural resources.

The prior appropriation doctrine historically did not limit where water was used. The few states that have enacted special laws to limit water transfers from one watershed to another provide a specific type of public interest review that focuses on the area where water originates. Such restrictions apply to new appropriations as well as to transfers of existing rights.

California's laws are illustrative. Because of its population distribution the state depends on moving huge quantities of water from sparsely populated areas that have copious water to growing cities where water demand is high. On paper, the legal protections for areas of origin in California are strong. For example, one state law gives an exporting area an absolute priority to the future use of the water over the priority of the importing area.[8] Another law reserves to the county of origin all of the water necessary for its future development.[9] As a practical matter, however, it would be difficult for an area or county of origin to cut off an urban area that has grown dependent on water imports.

Other states have attempted to protect areas of origin through a variety of legal mechanisms. Montana requires participation of the state in transfers of water out of a watershed; large transfers are limited and the state is obliged to consider public interest factors.[10] An Arizona law gives irrigation districts a veto over exports of water beyond their boundaries.[11] Colorado allows water conservancy districts to make transbasin diversions from the watershed of the Colorado River only if they will not inhibit or increase in cost the present or future water supply for the exporting area.[12] The Colorado law is interpreted to require districts that import water to the eastern side of the Rocky Mountains to construct special reservoirs for 'compensatory storage' in the watershed of the Colorado River.[13] However, there are no similar restrictions against large cities such as Denver that import the majority of the water from distant watersheds in Colorado.

State restrictions designed specifically to inhibit transfers of water beyond state borders raise constitutional problems. The United States Supreme Court has decreed that water is essentially an 'article of commerce', and restrictions that discriminate against interstate commerce violate the Commerce Clause of the United States Constitution.[14] To be constitutional, the regulation of water use must be impartial, treating equally users of water within and without the state.

Public trust doctrine

The examples described above require public interest review before a new water use or changed water use is approved. In some instances, however, courts have held that a state's decision to permit private use of public resources could be voided subsequently when water rights were allocated or transferred without review of the public interest. The landmark case involved California's Mono Lake.[15]

The public trust doctrine recognizes that water is fundamentally a public resource and that private interests in water should be advanced without inhibiting the public uses. The doctrine has ancient origins in civil and common law principles recognizing public servitudes such as the right of passage over navigable waters and the states' property rights in the beds of navigable waters. As applied, the public trust doctrine allows a court to reexamine established water rights in order to ensure that public values are protected, including the public value of environmental protection. The doctrine may be understood as a duty of continuing supervision over the granting and exercise of private rights to use water, but in practice it has not been interpreted so broadly except in the isolated California case involving Mono Lake.

Instream flow maintenance programs
Historically, western water law recognized only a limited number of water uses as worthy of legal protection. With very few exceptions (such as navigation and power generation), these 'beneficial uses' excluded values and activities that relied upon water remaining in the watercourse – now commonly referred to as instream flows. This excluded many uses that are today recognized as economically, ecologically, or socially important.

In recent years almost all of the western states have passed laws protecting instream flows (Gillilan and Brown, 1997). These states either allow state agencies to appropriate water rights to maintain streamflow levels or they remove from private appropriation the amount of water necessary to maintain desired flows. At present, only Arizona and Alaska permit individuals and private organizations to appropriate waters for instream flows. In all other states only a state agency can hold the right.

Statutory instream flow programs may not effectively protect streamflows because typically by the time a state appropriates rights for instream flows the water in the stream already has been fully appropriated by others. Relatively new instream flow appropriations will not prevent holders of senior water rights from using up all the water in the stream in dry years when a minimum flow is most needed.

Some states, however, authorize state agencies to buy or accept donations of water rights with priority dates sufficiently senior to maintain streamflows all or most of the time. Because all of the rights to use streams in most western states were claimed long ago, effective streamflow protection will depend on the acquisition of senior water rights. Private groups in some states, such as Oregon, Washington, and Colorado, have formed 'water trusts' that are authorized by state law to acquire senior water rights using private funds; these rights must be transferred to the state agency authorized to hold instream flow rights unless the state allows private entities to hold them.

Reserved rights for federal public lands

In much of the West, large expanses of land are owned by the federal government and managed for multiple public values. The government has reserved some of these lands for specific public purposes that require water: national forests, wildlife refuges, recreation areas, wilderness areas, and military bases. Although military bases require water for traditional consumptive uses, many of the other land designations require water for instream flow uses.

The doctrine of federal reserved water rights says that the federal government, when setting aside lands for public purposes that require water, impliedly reserved water rights sufficient to fulfill those purposes.[16] This court-made doctrine traces to precedents dealing with the establishment of Indian reservations and its development is discussed in the following chapter. In the context of federal public lands, however, the United States Supreme Court has read the doctrine restrictively to limit rights to the minimum amount of water necessary to accomplish the explicitly articulated federal purposes of each reservation.[17] It has also construed legislation creating federal reservations narrowly. For instance, national forests do not have reserved water rights for the instream flows needed for fish and wildlife because they were created primarily to provide a supply of timber.[18] In any event, the United States government has rarely taken action to enforce reserved rights, even where it has been found to have such rights.

Environmental regulation

Environmental laws enacted in the 1970s have established national programs and standards for water quality, wetlands, and endangered species protection. While they do not explicitly seek to override state primacy in water management, their enforcement indirectly addresses the effects of water allocation, development, use, and transfer. Thus, federalism concerns arise when they conflict with or curtail the uses of water under state water law. Several of the most relevant federal laws are discussed below.

Environmental impact assessment requirements   The National Environmental Policy Act of 1969 (NEPA) requires the assessment of potential environmental impacts of proposed 'major federal actions'.[19] After a public participation process, an agency completing this analysis explains its decision in a document known as an environmental impact statement.[20] NEPA applies to proposals that require a federal approval or license to use water in federal facilities where there will be a significant environmental impact. A few western states, including California, Montana, and Washington, have adopted laws with similar requirements for projects permitted or sponsored by the state.

Laws requiring an assessment of environmental impacts are important mechanisms for evaluating the effects of water development and transfer. The information developed in this process, presumably, provides a fair and comprehensive review of the public interest. Although courts have interpreted NEPA as essentially a procedural requirement (that is, it does not require an environmentally benign final decision), the mandated public scrutiny may reveal more efficient, less environmentally harmful approaches for achieving original project goals.

Clean Water Act   Water quality can decline with excessive depletions of a watercourse because the contaminants become more concentrated in the remaining flows. With some exceptions, however, state water agencies exclusively consider issues related to the quantity being allocated and not the quality.

Generally water quality is protected by the Clean Water Act (CWA).[21] Although the CWA is a federal law, most states and some Indian tribes have chosen to assume responsibility for its administration. Under the CWA, anyone who makes a 'point source' discharge of pollutants (such as from a pipe or ditch) into a waterway must have a permit that limits the quantity of particular pollutants according to standards established by the federal government.[22] The permit also must require sufficient limitations on discharges to protect the overall quality of the watercourse receiving the wastewater. States set the standards for water quality that are specific to particular waterways. The permitting program has effectively regulated industries and municipal sewage treatment plants that discharge wastes into rivers and lakes.

The CWA does not deal with declines in water quality caused by non-point source discharges such as run-off and drainage. Provisions in the Act encourage states to take action to control non-point sources and require that they identify waterways where the water quality is not effectively controlled by point source regulation. The states are then to impose 'total maximum daily loads' of pollutants that can be discharged into these water bodies. Lacking any firm enforcement mechanisms and surrounded by political criticism, this part of the program had not been fully implemented. The nation still has neither an effective program to prevent non-point source pollution nor any formal controls of water depletions to protect water quality.

A state agency theoretically could consider the potential effects of depletions on water quality resulting from new diversions or transfers away from the stream as part of the process of public interest review, but this is rarely done because states typically separate their administration of water allocation and water quality (Getches et. al., 1991). More typically, water laws protect the right to use a quantity of water even if it causes deterioration in water quality. When state water allocation laws come in conflict with federal water quality laws, the right to use water is generally considered superior to the protection of water quality.

Riverbed and wetland protection   A special program under section 404 of the Clean Water Act regulates 'dredging and filling' of 'navigable waters'. The statute defines navigable waters as all 'waters of the United States'. This has been interpreted administratively to include all adjacent wetlands, defined as areas capable of sustaining riparian vegetation. The activities covered are more than traditional dredge and fill operations undertaken to deepen channels for navigation. Depositing 'fill material' can include any construction in a waterway or wetland. Thus, the statute covers water projects, dams, and diversion structures.

The impact of section 404 on water development is much greater than pollutant discharge regulations. Almost any type of construction activity to develop or use water occurs in or on the banks of a stream or technically interferes with wetlands. Where streams or wetlands are affected by water development activity, section 404 requires the United States Army Corps of Engineers to conduct a global review of the public interest. In practice, the authority exercised by the Corps of Engineers is not nearly as broad as its powers, although the potential scope of its public interest inquiry is great.

Endangered species protection   The Endangered Species Act of 1973 (ESA) is another federal statute that can affect proposals to divert, develop, or transfer water.[23] The ESA absolutely prohibits any action by the federal government that would jeopardize the continued existence of an endangered species. Federal agencies considering activities that could jeopardize endangered species are required by section 7 of the ESA to consult with the United States Fish and Wildlife Service – or, for marine species, the National Marine Fisheries Service – to determine the effects of the development or action on the habitat of any endangered species. If, in the opinion of the Fish and Wildlife Service (or the National Marine Fisheries Service), the action would jeopardize the endangered species, the action cannot go forward unless there is a reasonable and prudent alternative that will not cause the jeopardy.

The ESA is extremely powerful because nearly every major water project – not just those undertaken directly by the federal government – requires some kind of federal approval (such as under section 404 of the Clean Water Act), or receives federal financing. Thus, the ESA has proved to be a formidable barrier to water development that could destroy fish or wildlife habitat where endangered species are found. The ESA, indeed, may be the most significant law affecting new water development.

Another section of the statute, section 9, prohibits actions that 'take' or 'harass' an endangered species. These terms are broadly interpreted to include harm to the habitats of endangered species. Unlike section 7, which is specific to federal agency actions, section 9 extends to private actions. The section has rarely been applied to private water development or uses. In one exceptional case, however, the Anderson–Cottonwood Irrigation District killed several Chinook salmon (an endangered species) while operating its pump diversion facility. The state court enjoined the irrigation district's activities, prohibiting it from 'taking' the endangered species.[24]

Federal Power Act For many years, federal law has required licenses for hydroelectric-generating dams located on navigable waterways or their tributaries. Congress enacted the Federal Power Act of 1920[25] to promote the coordinated development of rivers, and established a licensing agency known as the Federal Energy Regulatory Commission (FERC) to grant licenses for uses 'best adapted to a comprehensive plan' for each river.[26] FERC considers a variety of issues related to economics and other subjects. Although the Federal Power Act contains a provision saying that nothing in the statute shall affect state water laws, and another requires that an applicant for a license must show that it has complied with state laws, states cannot prevent a dam under the jurisdiction of FERC from being licensed and constructed.[27] Thus, state water law is subordinate to FERC's licensing authority.

Especially relevant to protection of the public interest is the mandate in the Federal Power Act that FERC's planning for a river should take into account all 'beneficial public uses, including recreational purposes'.[28] This was held to include consideration of the effects of the dam on anadromous fish.

Not surprisingly, older projects were built with little concern for such uses as fish and wildlife or recreation and subsequently have proved to be highly detrimental to fish. Historically, FERC was primarily concerned with maximizing a river's potential for hydroelectric power development. The agency did try to prevent negative impacts on navigation, but other public interests were not serious obstacles to the construction of dams. In recent years, however, as the licenses for old dams expired, often after a term of 50 years, the Commission has been more mindful of other public interests. This is partly the result of amendments to the Act and partly because of the enactment of the Fish and Wildlife Coordination Act of 1934, which requires FERC to give 'equal consideration' to protection of fish and wildlife.[29]

In granting new licenses and when relicensing existing projects FERC now often requires dam owners to release water in amounts and at times needed for fish and wildlife and other environmental purposes. These requirements have brought FERC into conflict with state agencies in some cases. Typically, the conflict has been over whether FERC or the state that issues water rights for the project has the last word on how much water must be left in the stream or that must bypass the dam to protect fish habitat. In one major case, the state of California tried to impose more stringent requirements on a power company for the benefit of fish. The Supreme Court held that the preemptive force of the Federal Power Act left such matters within the exclusive jurisdiction of FERC, so that the less protective federal requirements would apply;[30] it is sufficient if FERC considers the recommendations of the state fish and wildlife agency.

FERC's reputation for environmental protection is mixed. Although it has a strong environmental mandate, it fights tenaciously to license dams that are subject to laxer requirements than a state would impose. There are, however, modern examples of the agency taking significant protective action. In some cases, state laws are not strict and FERC's requirements, such as requiring bypass flows for fish, are the only public interest requirements. In a few recent cases, the commission has considered requiring removal of dams subject to relicensing in order to restore a fishery. The Edwards Dam in Maine, for instance, was recently removed and two dams that have blocked salmon migration in the Elwha River basin in Washington are targeted for removal.

Wild and scenic river designation   Under the Wild and Scenic Rivers Act of 1968 rivers may be designated by Congress or by a state nomination process to be protected against future development that would impair their free-flowing character as it exists at the time of designation.[31] The three categories of designation are 'wild', 'scenic', and 'recreational', determined by the amount of development already in the river corridor at the time of designation, with the 'wild' designation applying to essentially undeveloped and roadless river stretches and the 'recreational' category covering rivers with substantial development.

The primary effect of designation is that the federal government under its various regulatory programs cannot authorize construction of water projects that obstruct the flow of the river. For instance, FERC cannot license new projects on these rivers. In addition, designation of a river effectively reserves a water right to the federal government to maintain the river's distinctive characteristics, preventing depletions that would impair the flows to the extent that the purpose of the designation would be defeated. Managers of adjacent federal lands must also take the designation into account when making decisions about allowable land uses.

Specific river restoration laws   In the past decade, Congress enacted several laws calling for large-scale restoration of river environments. The Grand Canyon Protection Act of 1992, for example, ordered the United States Bureau of Reclamation to change the way it operated the Glen Canyon Dam in order to improve the downstream riparian and aquatic habitats.[32] As a result of the Act, the Bureau conducted an experimental 'flood flow' in 1996 – a large release intended to mimic historical spring run-off conditions in which high water levels with heavy sediment loads restored beaches and revitalized backwater native fish-rearing habitats. The Act explicitly directed the Bureau to manage the Glen Canyon Dam to protect, mitigate, and improve the natural and cultural resources of the river downstream – a dramatic expansion of the project's purposes when compared with the original authorizing legislation. The federal managers currently are engaged in experiments aimed at building upon incomplete knowledge of the river ecosystem, an approach known as 'adaptive management'.

In another example of legislatively mandated habitat restoration, the Central Valley Project Improvement Act of 1992 directed the Secretary of the Interior to dedicate and manage annually 800 000 acre-feet of water from the Central Valley Project for the primary purpose of fish, wildlife, and habitat restoration in California's vast and fertile Sacramento–San Joaquin River Valley.[33] Although this water was classified as 'surplus', irrigators participating in the large federal project had enjoyed its use during dry years, and thus faced cutbacks as a result of the new emphasis on habitat restoration. The Act also required these water users to pay surcharges on irrigation water to finance environmental restoration. The law's enactment culminated a successful lobbying effort by a coalition of diverse interests: environmental groups, commercial and sport fishermen, duck hunters, waterfowl organizations, Native Americans, and urban and business interests.

Ad hoc negotiations and other collaborative processes
Dozens of small, watershed-based voluntary groups have been formed throughout the western United States. They demonstrate the potential for finding creative, collaborative solutions to resource management problems (Kenney et al., 2000). In some cases such initiatives have addressed diminished streamflows in popular fisheries by crafting voluntary agreements among water rights holders to change the timing of withdrawals; in return, fisheries proponents have agreed not to seek regulatory actions that might limit the exercise of the appropriators' water rights. Other watershed groups have improved local knowledge of resource problems and potential solutions, coordinated management efforts among diverse government and private landowners, and engaged in direct restoration work in streams and adjacent public and private lands.

A smaller number of larger, higher-profile, initiatives have tackled thorny management issues in major river basins or sub-basins. These groups typically emerge amidst large conflicts and are governed by formal agreements or even federal legislation. Participants represent major stakeholder interests – irrigated agriculture, environmentalists, and the commercial fishing industry, for example – and participate partly in order to maintain a strong position in any resulting negotiated settlement. Multiparty initiatives of this nature are currently underway in California's Sacramento-San Joaquin Delta region (called 'CALFED') and in the Platte River basin.

Collaborative approaches also emerge out of conflicts over proposed water projects. Sometimes third parties affected by a proposed water development persuade proponents to take voluntary action to protect the interests of the public. But, unless there is a public process provided for under the law, it is difficult to initiate these negotiations. When third parties have sufficient political or legal leverage (for example, the threat of a veto under the Endangered Species Act) proponents of development activity are more likely to participate in negotiations. The fundamental problem with relying on such ad hoc negotiated resolutions is that the results are a function of the political power of the objectors. The results, then, are not consistent among similar projects and often provide incomplete relief where the objectors lack political or legal strength.

At the least, collaborative groups provide an opportunity for better expression of public interests. At best, they offer a forum in which otherwise marginalized 'third party' interests become integral to the crafting of creative solutions, and build relationships that may prevent some conflicts in the future.

Conclusion

Taken together, the several mechanisms for protection of the public interest in western water create rather uneven results. In some cases there is ample protection of fish and wildlife, but no consideration for recreation. States generally provide little protection for public values under prevailing water laws, so that the only effective protection is under federal laws. In other places state and federal agencies compete for control of water projects, not necessarily to ensure greater protection but to preserve their relative scope of jurisdiction.

Both the public and water developers would enjoy greater predictability if each state established a dynamic and comprehensive system of water planning covering issues such as impacts on rural communities, environmental effects, flood control, and protection of fisheries, wetlands, recreation, and drinking water. Only a few states have this type of water plan. Such a plan can provide standards that enable the decision maker under almost any of the programs now in place to judge a proposal more wisely and fairly. A comprehensive plan would identify a panoply of values and interests that could be affected by the development, transfer, or use of water. The plan could also discuss the relative importance to society of the values and their impacts.

Experts in the field of water law and policy should pursue methods to reduce the costs and increase the benefits of water use by providing broader and more effective protection of the public against negative effects of water decisions. Western states currently provide some protection through a combination of administrative review designed to protect the interests of the public, laws to protect instream flows, and environmental regulation. Protection could be further improved by establishing a comprehensive planning process to develop water policy that is more coherent and predictable.

2. INTERNATIONAL CASE STUDIES

Introduction and Overview

The western United States model of water resource management shares a fundamental principle with several other countries profiled below: water is a public resource, the use of which is permitted and supervised by the state acting as a trustee of the public good. As described in more detail in the following case studies of Australia, South Africa, and China, in recent years these countries have asserted this authority strongly and more directly than the western states of the United States, and have placed substantial restrictions on private rights to use water.

Both Australia and South Africa derived their legal systems from English models. While Australia rejected the English riparian doctrine as unsuited for its arid landscape, South Africa adopted a racially discriminatory version of riparianism favoring white landowners. Recently, with the advent of democracy in South Africa, the government rejected the riparian system and instead embraced a water use licensing system built upon a strong public trust doctrine. Australia also recognizes private rights to use water through limited licenses, which are subject to review and revision in the context of increasingly comprehensive resource planning. Australia's model is similar to the western United States in the primacy that states have in water resource management; like the United States however, the national government is playing an increasingly important role, particularly in environmental protection. South Africa's water law is national in scope and implementation, and seeks above all to redress past social wrongs.

China represents a different approach to water resource management, with highly centralized planning and priority placed on structural solutions to water quality and distribution problems. Its emphasis on rational, river basin engineering is reminiscent of the United States dam-building era dominated by the Bureau of Reclamation and Army Corps of Engineers, which drew to a close in the late 1970s. While water is fundamentally a public resource in China, the country is moving toward recognition of private rights in an effort to encourage limited market transactions in water rights.

Unlike the United States, the countries described below charge private users for the right to extract water from natural waterbodies and put it to use. In no case, however, has any of the countries yet attempted to include broader societal or ecological costs in the calculation of use charges.

Australia

Australia's Commonwealth Parliament has powers to legislate only in areas specifically designated in the Constitution. States hold the power to legislate in all other areas, and have traditionally assumed primacy over natural resources management. In recent years, however, the Commonwealth government's powers over environment and natural resources have been interpreted more broadly, based on its authority to enter into and enforce international agreements such as the United Nations Convention for the Protection of the World Cultural and Natural Heritage, the Convention on Wetlands (Ramsar Convention), the United Nations Convention on Biological Diversity, and bilateral migratory bird treaties with China and Japan. Sometimes the two levels of authority merge through cooperation: the Murray–Darling Basin of southeastern Australia is managed in a Commonwealth–state partnership, under which a basin council determines major policy issues concerning use of the basin's water, land, and other environmental resources. States are responsible for implementing the council's decisions.

As was the case in the western United States, Australia rejected the English doctrine of riparian water rights in the late 19th century as unsuitable for an arid land with substantial demand for irrigated agriculture. Instead, each of the Australian states adopted an administrative system for allocating water through licenses for specified periods of use. The state laws typically attached water rights to specific lands so that water rights automatically transferred with land ownership, and the amount of water used was legally controlled by restricting irrigated land areas, rather than limiting water quantities applied. The state also charged users only a small part of the costs of providing water and did not automatically terminate water allocations if users failed to make use of them. Similar to prior appropriation, the Australian system encouraged overallocation of water sources and did not address environmental or ecosystem concerns.

Australia's limited-term license contrasts with the appropriative water right in the western United States, which continues indefinitely so long as the user demonstrates a beneficial use of the water. It also differs from the prior appropriation doctrine's security for senior water users; instead, those with allocations under the Australian administrative system typically 'share the pain' in times of drought. For instance, in certain valleys north of the Murray–

Darling Basin where water is overallocated, irrigators receive their full allocation in only about 35 per cent of the years.

In another important departure from western United States water law, Australian states have recently added provisions to implement a national objective of ecologically sustainable development, aimed at:

- enhancing individual and community well-being and welfare by following a path of economic development that safeguards the welfare of future generations;
- providing for equity within and between generations;
- protecting biological diversity and maintaining essential ecological processes and life support systems.

An important component of ecologically sustainable development is the 'precautionary principle', which provides that if there are threats of serious or irreversible environmental damage, lack of full scientific certainty should not justify postponing measures to prevent environmental damage. In addition, traditional divisions are beginning to break down between natural resources legislation addressing water quantity issues and more recent environmental legislation dealing with water quality. Associated with this trend is a movement away from ad hoc regulation of individual project proposals to an increasing focus on planning that addresses water uses and environmental concerns across entire catchments, or watersheds.

Several states are moving away from water allocation systems that make ad hoc grants of water licenses that have led to significant overcommitment of the resource in many areas. Instead, they are shifting to planning processes to determine the overall quantity of water available for extraction by irrigators from particular water resources, and the times and other circumstances under which water is available. Approaches to water planning in Australia vary, ranging from highly centralized (Queensland) to broadly inclusive consensus processes (New South Wales). South Australia has delegated planning responsibility to regional water catchment boards, which enjoy considerable autonomy, although their plans must be approved by the state and comply with the State Water Plan. All the planning bodies are engaged in similar work: determining the quantity of water available for extraction by irrigators from particular water sources and the conditions by which it will be available (timing, and so on), and setting the broad parameters within which license allocations are managed. Current planning priorities are focused on stressed or environmentally sensitive rivers.

The planning processes now underway in Australia regularly encounter the difficulty of balancing values related to river and ecosystem health with those arising from consumptive water uses. In the context of water sharing, for example, must water first be set aside to satisfy ecosystem requirements before determining how much is available for extraction? The states vary in their responses. At one extreme, Victoria's Water Act does not even mention ecosystems in its definition of purposes, but rather makes an unequivocal commitment 'to continue in existence and protect' all existing public and private rights to water. The New South Wales Water Management Act gives priority to providing water to satisfy ecological requirements, but the water management plans now in development in that state do not appear to be implementing this priority, in part due to the dearth of scientific information on which to base assessments of ecosystem requirements. This challenge of scientific uncertainty is addressed in part by the precautionary principle, mentioned above as part of Australia's commitment to ecologically sustainable development. In practice, however, Australia's states have not yet actually shifted the burden of proof to those proposing water extraction to address any credible threats to the environment.

Attempts to deal with scientific uncertainty through adaptive management have run into difficulties with water users' expectations that their allotments will continue and a belief that the state must compensate them for any diminution of 'property rights' in water deliveries. Several states take the position that consumptive users are not guaranteed that their rights will not be curtailed if the plan changes. David Farrier (2002) describes how water may legally be 'clawed back' from existing users to satisfy ecosystem needs identified as a result of new scientific research after the plan is adopted. There is no provision in Australian law for compensation for 'takings' by state governments, and a recent review concluded that compensation for reductions in entitlements would not be justified if they resulted from a full-scale review carried out within an open and consultative planning process. Several states provide for limited instances of compensation, however, and others are examining opportunities to apply forward-looking subsidies to help water users to adopt efficient technologies that enhance environmental flows.

The Murray–Darling Basin has become a model for illustrating creative approaches to resolving the tensions confronted when water must be reallocated from existing uses to satisfy ecological and other demands. The Murray–Darling Basin Initiative is the largest watershed management program in the world. The watersheds of the Murray and Darling rivers drain an area of over one million square kilometers that sprawls over most of southeastern Australia. Five state governments, the national government, and community interests including irrigators and environmentalists joined in a partnership to carry out the 1992 Murray–Darling Basin Agreement that was negotiated 'to promote and coordinate effective planning and management for the equitable, efficient and sustainable use of the water, land and other environmental resources of the Murray–Darling Basin'.

Agricultural water use in the Murray–Darling Basin has a long history. Farming produces $10 billion annually. As David Farrier (2002) explains, water use under licenses (similar to water rights) granted by the state governments had expanded to the point that it was causing grave environmental problems – water quality degradation impacting urban areas, irrigation salinity caused by waterlogged soils, dryland salinity caused by loss of vegetation, depleted flows causing destruction of ecological resources, loss of more than 50 per cent of wetlands, and declines in biological diversity with 20 species extinct and another 16 endangered.

The Murray–Darling Basin Initiative has resulted in reductions in water use, a scheme of water charges that suppresses demand, and dedication of water to ecological purposes including instream flows. Remarkably, the council for the basin agreed that a balance had to be struck between consumptive and instream uses of water and placed a cap on water diversions from the basin's rivers. This means that any new uses must rely on transfers of water from existing uses, largely through a system of water markets facilitated by the initiative. A rather robust system of exchanges is described by Henning Bjornlund (2002). He also points out that the principle of pricing water at a level of full cost recovery works, showing that there is a high correlation between pricing and water demand because irrigators follow price signals.

One of the great successes of the Murray–Darling experience is that water has been reallocated from consumptive uses to address the array of environmental problems. According to Don Blackmore (2002), Chief Executive for the Murray–Darling Commission, the parties, while understandably not eager to suffer limitations on their water use, have generally approached the process in a cooperative spirit of civic responsibility.

South Africa

In 1998 South Africa's National Water Act completely overhauled the regulatory regime governing water resource management by replacing a racially discriminatory private rights system of water allocation with a system based on public rights. The new system applies the public trust doctrine to express the government's obligation to fulfill essential constitutional duties: equitable access to water; environmental protection and sustainable resource use; justifiable social and economic development; and effective recognition of the country's international obligations. South Africa's version of the public trust doctrine was based on the doctrine as articulated in the United States, according to lawyer Robyn Stein, who was instrumental in formulation of the new law (Stein, 2002). In fact, the doctrine as implemented in South Africa goes much farther in ensuring protection of public values than its American antecedent.

Fundamental to South Africa's water resource management is a constitutional guarantee that every citizen has a right of access to sufficient water. Furthermore, everyone is guaranteed a right to an environment that is not harmful to human health or well-being, and to have the environment protected for the benefit of present and future generations. These constitutional rights place affirmative obligations on the government to protect the environment and to meet basic human needs.

The National Water Act seeks to achieve the following objectives: meeting basic human needs of present and future generations; promoting equitable access to water; redressing the results of past racial and gender discrimination; promoting efficient, sustainable, and beneficial uses of water in the public interest; facilitating social and economic development; providing for growing demands for water use; protecting aquatic and associated ecosystems and their biological diversity; and reducing and preventing pollution and degradation of water resources.

In developing a new water policy, South African officials reviewed the application of the public trust doctrine in the United States, including the California Supreme Court's formulation of the doctrine in the Mono Lake litigation.[34] As written, however, the National Water Act took the doctrine much farther, defining the national government as the 'public trustee' of the nation's water resources and charging the Minister of Water Affairs and Forestry with ensuring that water is allocated equitably and used beneficially in the public interest, while promoting environmental values. Moreover, the National Water Act seeks to give effect to the development of participatory democracy in South Africa by mandating public comment and consultation at many stages of planning and decision making, and by requiring that information be easily accessible to the public.

The National Water Act abolished the riparian doctrine that emerged and developed in South Africa common law largely as a result of the influence of English common law. Instead, it treats water resources 'not only as ordinary property subject to the rules and assumptions of the private property system, but also as elements of the community's capital stock, the use and protection of which could affect the fate of the community' (Stein, 2002). Water use rights are viewed as *res comunis omnium* – property that is common to all and is not capable of private ownership.

Private water use is regulated through a licensing system, which makes allowance for allocation decisions that are sensitive to equity considerations and local and regional demands. All water use licenses are subject to conditions and are issued either for a fixed period (which may not exceed 40 years) or for a limited extension period. A license is not required for domestic uses, small gardening, and livestock watering. The Act provides for water use charges in order to support the implementation of statutory objectives. The pricing strategy differentiates between different water users on the basis of the extent of their water use, the quantity of water they return to the source, or their economic circumstances.

Clearly, the new water regime in South Africa made tremendous changes in the way water was allocated and managed. Many existing water users had to adjust to new limitations on uses to which they previously enjoyed a legal right. The National Water Act provides for compensation to be paid where existing entitlements to use of water are lost through the refusal of a license application or the grant of a lesser use than the original entitlement, but only when such loss constitutes destruction of or severe prejudice to the economic viability of an undertaking in which the water could have been beneficially used. In order to meet constitutional duties to protect human well-being and environmental health, the National Water Act provides for a 'Reserve', defined as 'that quantity and quality of water required to (a) satisfy basic human needs for all people who are, or may be supplied from a relevant water resource; and (b) protect aquatic ecosystems in order to ensure ecologically sustainable water development and use'. The amount of water in this Reserve is as yet uncalculated.

Additional provisions in the National Water Act include a resource quality protection system based on classification of all significant water resources, relatively stringent pollution prevention measures applying 'polluter pays', and a requirement that water resource decisions take into account international obligations.

China

China is a densely populated country with water resources distributed sometimes far from population centers. Thus, government planning and priorities have aimed at moving water over long distances to satisfy human needs. More recently, government policies have sought to repair unbalanced river ecosystems as well. Like the other countries studied, China is confronting problems of how to coordinate expanding water demands for consumptive uses with environmental protection.

The Chinese government began to regulate and manage rivers through water resources legislation in the 1980s. In the 1990s, China established sustainable water use as a key strategy for economic and social development. Planning follows a model of rational exploitation, aimed at distributing water resources between different river basins and within river basins (Xi et al., 2002). Allocation between river basins is accomplished by transbasin projects that import water to regions where human demands are high and streamflows are low. Water use is allocated within basins by controlling water quantity and quality to enhance water management and improve water use efficiency.

China's constitution provides that water is a public resource; it is owned either by the state or by collectives. The state governs the use of public resources by units and individuals. A permit is required to withdraw water for use, except for small quantities of water exempted from the permitting process for specified purposes such as family living and livestock watering. Permits are issued based on categories of use, with the highest priority for domestic use. The law provides for charging fees for water use to support a fund for water resources conservation. Water permits may be conditioned to ensure that ecological demands for water are met first. As yet, there is no effective market mechanism for trading in water permits.

The Chinese government uses a system of water function zoning to manage water resources. The system uses the river basin as the zoning unit, and seeks to maintain existing functions within each zone. Surface water is categorized into five classes according to the purpose of use and protective targets for water quality. Total volume is then established based on two types of water use: economic and ecological. At present, agriculture accounts for about two-thirds of total national water consumption; industry uses about one-fifth; and about one-tenth of the total consumption is for subsistence use. Water function zoning is aimed at rational development and effective protection of water resources, and is based upon national economic and social development plans.

According to Professor Wang Xi of Wuhan University, the National People's Congress is considering amendments to the Water Law to deal with several shortcomings in the present system (Xi et al., 2002). These include lack of coordination between water resource managers in the provinces and local governments, inconsistent water quality standards, lack of regulation of agricultural water users and generally inefficient irrigation practices, lack of connection between water resource and water quality regulation, unclear responsibilities for basin management authorities, ecological impacts when water is diverted from one river basin to another, and lack of clarity in the attributes of a water right. One of the greatest problems is the underpricing of water. Cheap water provides no incentive to conserve. This, in turn, tends to deplete streams and deprive ecosystems of water. Professor Xi believes that it is vitally important to address this issue and to institute a market-based system of water trading under government supervision if China is to meet the demands of a rapidly expanding economy with a large population while ensuring environmental protection (Xi et al., 2002). In the past, the government bargained away a public asset in water contract negotiations that did not include members of the public – including those who needed service and those who would have to pay the new tariffs. The public had no notice of or role in setting the terms of the contract yet they were profoundly affected by the outcome. Most people in the area who had customarily paid little or nothing for water suddenly discovered that they would have to pay a substantial portion of their income or lose water service. The entire process was oblivious to the customs, economic status, or views of the people who were most affected by the decision.

3.   LESSONS LEARNED: ANALYSIS AND CONCLUSIONS

Introduction and Overview

All water laws are designed to fit a particular social and economic milieu. Perhaps more than other legal regimes, water laws in the United States have been tied to the needs of an era. Western water law was created amidst a nineteenth-century national expansion effort. Expansion depended on generating wealth in the undeveloped West and settling wild and barren lands. The enterprises that held promise were mining and agriculture, and both required water. To allocate and use that water efficiently, a simple, locally administered system was needed. It made sense to adopt the prior appropriation system, allowing anyone who had a productive use for water to claim a 'right' to use it. Although the federal government then owned most of the land, the job of defining water law was left to the states. Local user groups and later the states generally adopted the commonsense 'first-in-time' rule and enforced rights of the water users.

Since the prior appropriation system was created, the conditions that supplied a rationale for it have radically changed. The West has been settled. Mining and agricultural have become relatively minor factors in the economy. Today, regional and national values support protecting water in its natural state. Although water law in the western states has become much more complex than it was in the original incarnation of the prior appropriation doctrine, it has been difficult to give full meaning to environmental and public values because of the existence of property rights and the perceived equities of those who hold rights.

By looking at the experiences of other nations, the United States can test new ideas that may be adaptable to its own conditions and culture. The primary lessons are that, notwithstanding profoundly different histories, people in all of the nations studied, as in the United States, are dealing with changing public values and situations in which responding to those changes causes controversy. The experiences of Australia, South Africa, and China teach that there is broad public concern with the results of water law decisions. The United States is not alone in its efforts to integrate environmental protection and to deal fairly with the expectations of water users. It does seem, however, that water policy changes in the United States come more slowly and encounter greater resistance. Change, when it does come, appears to be propelled by looking outside traditional institutions to the grassroots public or to norms set by other governments, factors that have been influential in other countries as well.

Environmental Concerns Are Integral to Water Resource Management

All the countries profiled in the case studies share with the western United States a growing concern for water's ecological values. Australia and South Africa have, on paper at least, articulated environmental protection as a fundamental goal of water resource management. None of the western states in the United States has made such a strong commitment, despite consistent expressions of public support for environmental protection.

The United States approach, with its strong emphasis on property rights, has struggled with a new overlay of (mostly federal) environmental regulations, some of which undermine the security of private water users' historical rights. By contrast, the more comprehensive, deliberative approaches to watershed or river basin planning demonstrated in Australia, South Africa, and China suggest that public values might best be served when historical uses are periodically re-examined in the context of current ecological, social, and economic conditions.

Scientific understanding of environmental conditions is ever changing. Given this reality, Australia's commitment to the precautionary principle offers an interesting model for dealing with uncertainty while allowing decision making to move ahead. In the United States, by contrast, environmental regulation usually applies once harm has already occurred or is imminent. For example, plant and animal species must be in danger of extinction before they are considered for listing under the Endangered Species Act, an action necessary to trigger legal protection required for their recovery. The Superfund law (CERCLA)[35] focuses on cleaning up old hazardous waste sites, many of which have destroyed aquifers. The Resources Conservation and Recovery Act (RCRA)[36] includes a well-head protection program to prevent contaminants from seeping into wells, but most of the Act is, like the Superfund law, applied to force the clean-up of aquifers and other resources damaged by waste disposal.

Some of the federal environmental laws have 'precautionary' elements. The Clean Water Act – the primary water pollution law in the United States – bans discharges of pollutants into United States waters. Yet it operates to permit discharges into the waters of the United States so long as they do not exceed thresholds set by federal law. Notably, many states – the entities charged with giving pollutant discharge permits – have laws that say that if the requirements of the Act conflict with laws allocating water rights, the water rights will prevail. The notion that water use rights are 'property' allows them to prevail even if the use might conflict with the Act's goals of protecting human health and the environment. It seems useful at least to reassess United States approaches in light of the philosophy of erring on the side of protection that is inherent in the precautionary principle.

Equity Concerns for Existing Water Users: Expectations and Changing Rules

The prior appropriation doctrine that developed in the western United States offered a special sense of security for water rights holders. Their rights of use were indefinite and without charge, their seniority was clear and enforceable against later users, and they were led to expect stream conditions to remain essentially unchanged from the day of first appropriation. Alas, as demonstrated many times in recent years, such security was a false hope. If people believed that nothing could change, they were sorely disappointed.

Water users today face new regulations that can limit historical use rights, as well as chronically uncertain climatic conditions. Rights have had to give way to changing community values. And there are diverse 'communities' whose values are now considered – national, local, tribal, scientific, recreational, and so on. When these values are embodied in regulations, the result can be to redistribute a finite stock of water and upset the settled expectations of those who counted on 'certainty' from the legal system. As historical uses are no longer viewed as sacrosanct and are increasingly viewed in the context of changing social and environmental concerns, water users resist change because it reduces their security and the predictability of their water supplies. They argue that it is unfair to expect them to shoulder the burden of shifts in broader public policies.

How much certainty should water users be able to expect? One of the greatest certainties, especially in semi-arid climates, is the *uncertainty* of nature – that streamflows will vary seasonally and annually. But even property rights – in land and other things – can be expected to change. Land, for instance, unlike water flows may not change physically but the rules for how it is used do change. Land use and environmental regulations profoundly affect what owners can do with the land they own, and these property owners have grown to expect, if not enjoy, the reality that as social and economic conditions change, so will the nature of their property rights. So the question becomes: What level of certainty is it reasonable for owners of water rights (like owners of land) to expect?

This conflict is not unique to the United States, although the water laws of other countries described here did not seem to encourage the degree of confidence among water users that the prior appropriation doctrine did in the

United States. South Africa reformed its water law to effect a vast redistribution of rights to use water from a few large landowners so that the majority of the population could have access to opportunities to achieve economic viability in farming and access to basic water supplies. Although the fundamental assumptions of the earlier system were deemed morally unsound, the new water law nevertheless acknowledged the possibility of economic harm justifying compensation in the transition to the new water code that authorizes payments for the most severe consequences amounting to expropriation.

In Australia, measures to reallocate – 'claw back' – water from existing uses to satisfy environmental or other new demands provoke demands for compensation for economic harms. As described in the case study, this issue is not yet resolved. One suggested approach is to subsidize efficiency improvements in agriculture. This invites comparison with similar United States initiatives such as the Central Valley Project Improvement Act where federally funded irrigation delivery improvements in California are aimed at gaining additional water for wildlife and wetlands.

In the United States, compensation has taken several forms. As described above, the state of Colorado enacted area-of-origin protection measures that require water developers in the Colorado River Basin to ensure that the communities near the water source have sufficient water for future needs. This has been accomplished primarily through 'compensatory storage' – constructing new water storage projects located where the basin of origin speculates they may be needed someday.

Without basin-of-origin protection legislation, non-governmental conservation groups have used financial incentives to encourage water users to provide environmental protection or even to convert their diversion rights to instream flows. Water users who are denied all economic benefit of their water caused by government regulation will argue that they have suffered a "taking" of property rights as defined by the United States Constitution and are entitled to compensation. In practice, such claims have almost never succeeded. It is extremely difficult to establish that water users reasonably expected that the ability to use water would never be affected as social and economic conditions change over time. Although the claims of these users may not amount to legal rights to compensation they nonetheless raise equitable concerns. Therefore, it is highly desirable to smooth and expedite a transition in the way water is used through voluntary payments to established water users by governments or private, non-governmental organizations.

Seeking a Balanced Use of Markets and Regulation

Market mechanisms – properly integrated in a regulatory system – can be vital tools in protecting public values, and charging prices that reflect the true value of water can encourage conservation. By making water rights transferable, water can be moved to higher-value uses, resulting in greater efficiency. And compensation for loss of water rights (or expectations concerning the nature of those rights) can ease the transition to a sustainable system that incorporates public values. Third party impacts of water trading, such as changed return flows and environmental damage, can be partially accounted for in prices (Etchells et al., 2002). The role of markets is discussed in more detail in Chapter 2. But while the benefits of market economics cannot be denied, there is a growing consensus that markets must be tempered by rules to protect people and public values that are not well represented in the market.

In Australia, full-cost water pricing has been a tool in bringing about reductions in water use within the Murray–Darling Basin. Likewise, a system of water trading has moved water from the lowest-valued crops to more lucrative types. It seems clear that the trading system would not have been invented nor would water prices have increased had it not been for a rather drastic commitment to cap water consumption. National legal provisions drove these innovations. Although the parties to the Murray–Darling Initiative came together voluntarily, it was based on a widespread understanding that existing uses must be curbed.

China, where the market system has a shorter history, now recognizes the importance of water pricing as an incentive to conserve and thus a significant means of keeping water in rivers for ecological purposes. Experts also urge wider use of both pricing and water trading.

To be sure, the benefits of market mechanisms can be eclipsed by disadvantages. The disregard of third-party interests and values that are considered 'market failure' by economists can be largely avoided by a regulatory system that affords them respect and representation. As Miguel Solanes (2002, p. 4) argues, it is the role of a water rights system to balance public and private values because 'water is not an ordinary commodity. The peculiar characteristics of water resources stem from its multiple environmental, economic and social roles'. Specifically, the water rights system must include the means to ensure sustainability, protect third parties, preserve the environment, and advance the public interest. Market mechanisms can be mobilized to these ends, but they alone will not suffice without sideboards, standards, and limits. The fiasco of failed privatization of water utilities in places like Cochabamba, Bolivia, is testimony to the need for governmental stewardship of public values to guide the use of market forces in allocating water.[37]

'Outside' Influences and Water Policy Reform

In the western United States, the concept of state primacy over water resource management is undergoing change as the influence of federal environmental laws and local citizens encroaches on the hegemony of state agencies. The growth of citizen involvement has encouraged change in other countries as well. 'Higher law' in the United States has meant federal legislation, but there is an analog in other countries: the influence of international agreements and norms.

In Australia, the national government used its authority to implement international agreements to incorporate new environmental review procedures. This resulted in curbing presumed rights to continue traditional approaches to managing private water use in order to comply with these international norms. Australia has a two-tiered, federal-type system like the United States. It has had less state–national conflict, however, as the Australian states have more willingly incorporated national norms.

South African water law is also influenced by international law, having cited it as part of the effort to reform the apartheid system. It acknowledges international commitments, especially related to human rights. China also incorporates agreements with other countries into its water resources planning.

In the United States, with the exception of a few treaties allocating rights to use transboundary waters, the influence of international law is insignificant. Instead, federal environmental laws provide the critical 'outside' influence on western states. The Environmental Protection Agency, United States Fish and Wildlife Service, National Marine Fisheries Service, Army Corps of Engineers, and Federal Energy Regulatory Commission all are now regular players in public decisions regarding water development and management. The fit has been less than comfortable, however, with states resisting the perceived 'heavy hand' of federal involvement. Better incorporation of national standards in state water laws and administrative practices – looking, perhaps, at the state laws in Australia for examples – might reduce the number and intensity of federalism conflicts that regularly arise in western water management.

CONCLUSION

The observations we have made based on how water law operates in a few other countries reaffirm proposals for reforming water policy in the United States. The problems in each of the countries studied, and many of the obstacles to solving them, are not unique. Moreover, the approaches used in some countries may be worthy of experiment in the United States. In instances where environmental and other public interest factors are better reflected in the water decisions of another country, it may be appropriate to test those methods in the United States.

The importance of environmental protection has become an overriding concern in water allocation and management. Yet its integration into water decision making in the United States is awkward at best. Most US states have not fully integrated environmental factors into their water laws and policies and protection is achieved largely as a result of federal laws that trump water rights. A lesson from abroad is that confronting the issue directly and making environmental protection an essential role for water law may be preferable to the United States approach of putting environmental laws in tension with water laws.

All countries struggle to deal with the equities of how to expand the distribution of benefits from water use. But a major concern in the United States is how to prevent unfairness to those who have grown accustomed to receiving a disproportionate share of the resource. Reallocation of water is difficult and can seem unfair. Although there are few formidable legal barriers to reallocation in order to protect the public interest, using compensation can ease the process of change. The huge changes in Australia's Murray–Darling Basin and in South Africa show that change can come peacefully and fairly if supported by the larger community.

The market system has proved to be a highly effective tool in making water use more efficient in Australia as it has in the United States. Pricing is a part of South Africa's reform effort. China has underutilized pricing and market forces, but is being urged by experts to use these tools in the future. On the other hand, turning loose the unbridled market on the people of Cochabamba has illustrated the difficulty of liberalizing water policy too quickly and without regulatory measures to protect the public interest. As the United States tries to find the best use for economics in water policy, it is helpful to see that the idea is one that is being embraced all over the world to promote beneficial transfers and efficient water use. But markets in a vital resource like water must be carefully regulated to safeguard the public's interest in water.

Where water reform has occurred, it is usually the product of new institutions or pressure from outside traditional systems. Australia has recognized its international obligations to pursue sustainability and the national government has put pressure on states to deal with water to prevent ecological harm and waste. Although the Australian states most likely never would have made the reforms themselves, local people are cooperating because they recognize that the national policy is rational and ultimately will benefit the country. Collaborative processes in the United States also show that once the national government sets a broad policy, local representatives who formulate and carry out on-the-ground solutions can do the heavy lifting. But it does not appear that many significant changes in water policy are initiated within state institutions without external pressures.

NOTES

1. *Bonham v. Morgan*, 788 P.2d 497 (Utah 1989).
2. Nevada Revised Statutes § 533.370(3).
3. Wyoming Statutes § 41–4–503.
4. California Water Code § 109.
5. Idaho Statutes § 42–203A.
6. *Shokal v. Dunn*, 707 P.2d 441 (Idaho 1985).
7. *Shokal v. Dunn*, 707 P.2d 441 (Idaho 1985), citing *Young & Norton v. Hinderlider*, 15 N.M. 666, 110 P. 1045, 1050 (New Mexico. 1910).
8. California Water Code § 10505.
9. California Water Code § 10505.5.
10. Montana Code § 85–2–402(5).
11. Arizona Revised Statutes § 45–172(5).
12. Colorado Revised Statutes § 37–45–118(1)(b)(II).
13. *Colorado River Water Conservation District v. Municipal Subdistrict, Northern Colorado Water Conservancy District*, 610 P.2d. 81, 84 (Colorado 1979).
14. *Sporhase v. Nebraska*, 458 U.S. 941 (1982).
15. *National Audubon Society v. Superior Court*, 658 P.2d 704 (California 1983); *In re Water Use Permit Applications*, 9 P.3d 409 (Hawaii 2000).
16. *Arizona v. California*, 373 U.S. 546 (1963).
17. *Cappaert v. United States*, 426 U.S. 128 (1976).
18. *United States v. New Mexico*, 438 U.S. 696 (1978).
19. 42 U.S.C. §§ 4321–4370, 4321(2)(a)(1)).
20. 42 U.S.C. § 4332(2)(c).
21. 33 U.S.C. §§ 1271–1387.
22. 33 U.S.C. § 1362(14).
23. 16 U.S.C. §§ 1531–1543.
24. *Department of Fish and Game v. Anderson-Cottonwood Irrigation District*, 11 California Reporter 2d. 222 (California Appeals 1992)
25. 16 U.S.C. §§ 790 et. seq.
26. 16 U.S.C. § 803.
27. *First Iowa Hydro-Electric Cooperative v. Federal Power Commission*, 328 U.S. 152 (1946).
28. *Udall v. Federal Power Commission*, 387 U.S. 428 (1967), citing 16 U.S.C. 797(e).
29. 16 U.S.C. §§ 661–667, 661.
30. *California v. Federal Energy Regulatory Commission*, 495 U.S. 490 (1990).
31. 16 U.S.C. §§ 1271–1287.
32. 106 Stat. 4600.
33. 106 Stat. 4706–4731.
34. *National Audubon Society v. Superior Court of Alpine County* 658 P.2d 704 (California 1983).
35. Comprehensive Environmental Response, Compensation and Liability Act of 1980; 42 U.S.C. §§ 9601–9675.
36. 42 U.S.C. §§ 6901–6992k.
37. This observation is based on the remarks of Rocio Bustamante at the conference 'Allocating and Managing Water for a Sustainable Future: Lessons from Around the World', held at the Natural Resources Law Center, University of Colorado School of Law, 13 June 2002.

LITERATURE CITED

Bates, Sarah F., David H. Getches, Lawrence J. MacDonnell and Charles F. Wilkinson (1993), *Searching Out the Headwaters: Change and Rediscovery in Western Water Resource Policy*, Washington, DC: Island Press.

Bjornlund, Henning (2002), 'Water exchanges: Australian experiences', paper for the Natural Resources Law Center, University of Colorado School of Law, Boulder, CO.

Blackmore, Don (2002), 'Lessons in water allocation: roles for governments and markets', paper for the Natural Resources Law Center, University of Colorado School of Law, Boulder, CO.

Etchells, Teri, Hector Malano and Thomas A. McMahon (2002), 'Overcoming third party effects from water transfers', paper for the Natural Resources Law Center, University of Colorado School of Law, Boulder, CO.

Farrier, David (2002), 'Protecting environmental values in water resources in Australia', paper for the Natural Resources Law Center, University of Colorado School of Law, Boulder, CO.

Getches, David H., Lawrence J. MacDonnell and Teresa Rice (1991), *Controlling Water Use: The Unfinished Business of Water Quality Protection*, Boulder, CO: Natural Resources Law Center, University of Colorado School of Law

Gillilan, David M., and Thomas C. Brown (1997), *Instream Flow Protection: Seeking a Balance in Western Water Use*, Washington, DC: Island Press.

Glennon, Robert (2002), *Water Follies: Groundwater Pumping and the Fate of America's Fresh Waters*, Washington, DC: Island Press.

Kenney, Douglas S., Sean T. McAllister, William H. Caile and Jason S. Peckham (2000), *The New Watershed Source Book: A Directory and Review of Watershed Initiatives in the Western United States*, Boulder, CO: Natural Resources Law Center, University of Colorado School of Law.

Solanes, Miguel (2002), 'Water: rights, flexibility and governance: a balance that matters?', paper for the Natural Resources Law Center, University of Colorado School of Law, Boulder, CO.

Stein, Robyn (2002), 'Water law in a democratic South Africa: a country case study examining the introduction of a public rights system', paper for the Natural Resources Law Center, University of Colorado School of Law, Boulder, CO.

Xi, Wang, Zhang Xiaobo, Li Wenkai, Gu Dejin, Zhou Yanfang (2002), 'Managing water resources for a sustainable future: law, policy, and methodology of China', paper for the Natural Resources Law Center, University of Colorado School of Law, Boulder, CO.

4. Protecting indigenous rights and interests in water

David H. Getches and Sarah B. Van de Wetering

With assistance from David Farrier, University of Wollongong; Robyn Tanya Stein, Bowman Gilfillan Inc.; Wang Xi and colleagues, Wuhan University; and Marcos Terena, Coordinator General of Indigenous Rights, Mato Grosso do Sul

Access to water is fundamental to the right of indigenous people to use and enjoy their lands and maintain the integrity of their territories. Therefore, water rights are necessary for the cultures of tribes to survive in the 'melting pot' of the United States. Economic survival in arid environments often demands that indigenous communities have enough water for irrigation. In addition, many tribes still survive on fishing and hunting, which requires healthy rivers and lakes.

John Wiener (2002) has discussed the inextricable historical linkage of culture and livelihood. Where livelihood is dependent on the environment, as it is for native peoples who historically were hunters, fishers, and gatherers, environmental change has undermined indigenous cultures. As native peoples become alienated from an environment in which their culture has evolved, they experience abrupt and sometimes destructive change.

The nineteenth-century allotment policy in the United States, designed to separate Indians from their hunting and fishing lifestyle and confine them to small farms, is an apt illustration of this alienation. An explicit goal of the policy was destruction of communal ownership, itself at the heart of tribal cultures. The policy failed and was ultimately rejected, but not before 90 million acres of tribal land was lost (Getches et al., 1998). Along with the land went many cultural practices. This cultural loss was further attenuated by federal policies designed to eliminate native religion and political organization.

Now the challenge in the debate over indigenous water rights in the United States is to secure sufficient water to sustain cultures and economies. Today,

United States policy aims to promote economic and political self-determination while allowing for the survival of traditional cultures. For some tribes, economic well-being depends on agriculture; for other tribes it depends on fishing and hunting. All tribes need water for domestic purposes and some need it for commercial and industrial uses, such as mining, manufacturing, and tourist enterprises.

1. INDIGENOUS RIGHTS UNDER THE UNITED STATES FEDERAL SYSTEM

Protecting the rights of indigenous peoples when state governments make water decisions arguably should be part of advancing the public interest, as discussed in earlier chapters. But advancing the public interest is complicated because the 'public interest' furthered by satisfying national obligations to Indian tribes in the United States sometimes conflicts with the policies or asserted governmental authority of state governments. Although United States law recognizes certain sovereign rights and property interests of tribes and is 'supreme' under the United States Constitution, states may resist federal interference with their presumed autonomy in dealing with water. In this respect, the conflict between federal laws satisfying obligations to tribes and state laws protecting what they view as their resources is similar to the tension between federal enactments that promote the national environmental interest and state policies that often respond to more localized or short-term interests. Assertion of Indian rights by the United States government itself has been complicated by the nature of the federal system in the United States and at times also by the tribes' lack of influence in national politics. Nevertheless, an impressive body of law recognizes significant tribal rights to water and supports tribal control of water allocation and management within Indian reservations.

Indian Rights in Historical Context

The foundational principles in Indian law trace to the earliest days of the nation. These principles define a fiduciary relationship in which the national government is charged with protecting the rights of tribes as property owners and sovereigns. Property rights extend to certain lands, usually within reservations. Sovereign rights include control of those lands and activities on them. The Supreme Court has recognized that Congress wields broad powers to implement this obligation, but also to extinguish tribal rights when law makers determine that it is in the interest of the country to do so. This great federal power in the area of Indian affairs has been invoked frequently to limit states' efforts to encroach on the property rights and self-governing authority of tribes within their territory.

In modern times the federal policy toward Indian tribes has favored self-determination and economic self-sufficiency. However, this has not always been the case. In the late nineteenth century, for instance, national policy sought to assimilate Indians into the mainstream of society by breaking up communally held tribal lands and promoting individual farm cultivation. Whether the national goal was to promote individual ownership or collective self-sufficiency on the lands reserved for tribes and their members, access to sufficient amounts of water has always been essential.

The integrity of tribal land, water, and the natural world is often at the heart of traditional cultures and spiritual life. Often, this integrity has been violated by federal policy. Tribes of the Great Plains were placed on reservations and forced to give up their far-ranging hunts. In the desert Southwest, some tribes had established aboriginal irrigation cultures using the sparse and seasonal streams that were available. In the Northwest and Great Lakes regions reservations were created that limited the homelands and the historic fishing pursuits of native peoples. In each case, encroaching populations of non-Indians and the resulting competition for water and water-dependent resources threatened the ability of Indians to survive on their reservations. Nonetheless, national policies in the era of homesteading and westward expansion encouraged this settlement. The resulting establishment of non-Indian communities and creation of property rights in land and water have conflicted and competed with the Indians' capacity to use natural resources necessary for their survival.

As described in the following section, in the early days of the twentieth century, the US Supreme Court announced a remarkable doctrine of water rights that favored Indian tribes in their attempts to secure sufficient water to make their reservations useful.[1] The 'reserved rights doctrine' guaranteed tribes the right to use water to fulfill the purposes for which their reservations were established. The right could be exercised anytime in the future, even if non-Indians had used the water first and had been granted rights under state law.

In practice, the tribes' exercise of their ostensibly potent reserved water rights for their reservations has been problematic. The tribes have lacked capital to put their water rights to use and must now compete with non-Indians who have built their economies using the water to which the Indians are entitled. As a result, most tribes have remained in a state of poverty and reservations are largely undeveloped. Some tribes near population centers have sought economic development by legalizing gambling within their reservations, which is possible because their independent sovereignty makes them immune from state laws prohibiting gambling. But in most places long-term economic wellbeing and cultural survival for tribes on their reservations depend on tribes asserting and using their water rights for agricultural or industrial development.

Increasingly, tribes have pressed for a vindication of their theoretically great but actually underutilized water rights. Non-Indians are aware that the tribes pose a threat to their economic security. Because investments and property values are undermined by uncertainty, non-Indians and the western states that tend to support non-Indian interests have also urged that Indian water rights should be legally determined. Judicial processes, now underway in most states to determine Indian water rights, are lengthy and expensive. In recent years several tribes' water rights have been resolved in negotiated settlements and implemented through federal legislation. This remains the preferred method of quantifying tribal water rights, primarily because it infuses federal funding into solutions that enable tribes to use their water rights while protecting established non-Indian uses.

A Doctrine of Indian Water Rights

> Water is the blood of our tribes and if its life-giving flow is stopped, or it is polluted, all else will die and the many thousands of years of our communal existence will come to an end.
>
> Frank Tenorio, San Felipe Pueblo

Reserved rights

There are hundreds of Indian reservations in the western United States. Most of the reservations were created when the tribes who once hunted, fished, and gathered over large expanses of land signed treaties and agreements, often reluctantly or under pressure, to give up their claims to large territories in exchange for smaller reservations. Typically, the treaties and agreements said nothing about water rights.

As a general matter, all reservations were established to be permanent homelands for the tribes, where they could survive and be self-sufficient. Invariably, the reservations required water because on some reservations the government sought to convert the Indians into farmers in an arid region where agriculture is difficult without irrigation; on others, Indians were located along rivers to ensure that they could continue fishing to sustain their livelihoods and culture.

The fundamental legal principle giving rise to Indian water rights is stated simply: The establishment of a reservation results in the creation of a right to take water sufficient to fulfill the purpose of reserving the land for the Indians. In the words of the United States Supreme Court:

> The reservation was a part of a very much larger tract which the Indians had the right to occupy and use and which was adequate for the habits and wants of a

nomadic and uncivilized people. It was the policy of the Government, it was the desire of the Indians, to change those habits and to become pastoral and civilized people. If they should become such the original tract was too extensive, but a smaller tract would be inadequate without a change of conditions. The lands were arid and, without irrigation, were practically valueless. ... The Indians had command of the lands and the waters – command of all their beneficial use, whether kept for hunting, 'and grazing roving herds of stock', or turned to agriculture and the arts of civilization. Did they give up all this? Did they reduce the area of their occupation and give up the waters which make it valuable or adequate?[2]

The landmark 1908 case (*Winters* v. *United States*) in which the Supreme Court announced the doctrine of 'reserved water rights' arose on the Fort Belknap Indian Reservation in Montana where the Gros Ventre, Piegan, Blood, Blackfoot and River Crow Indians had been placed after a series of treaties. These agreements limited them to a small fraction of their former territory. It would have been grossly unfair to the Indians to confine them to a reservation without the means to eke out a living. So the reservation – and appurtenant water rights – had to be sufficient to meet the tribes' present and future needs.

Furthermore, the government plan involved dividing up the reservation lands into individual land holdings, allotting parcels of land to heads of Indian families to be cultivated, and then opening to non-Indian homesteaders the unallotted (or 'surplus') lands on the reservation as well as the off-reservation lands the tribe ceded. The Court recognized the government's intention to 'civilize' the Indians by making them individual farmers and breaking up the communally held tribal lands. Unless irrigation water was available to the Indians for their small farms, this grand scheme for 'civilizing the Indians' and distributing 'surplus lands' to white settlers might collapse. So, more than the welfare of the tribes depended on having adequate water for the reservations.

The reserved rights doctrine of *Winters* became the cardinal rule of Indian water rights. Later in *Arizona* v. *California* (1963), the doctrine was applied to federal lands reserved as parks, forests, military bases, and other public uses.[3] (See the discussion of federal reserved water rights in Chapter 3.) As with Indian lands, the quantity of water reserved depended on the purposes for which the reservation was established.

Over the years, the reserved rights doctrine has promised more than it has delivered. The government has rarely applied the doctrine in litigation to assert Indian rights as against non-Indian water users. Until about 30 years ago, most tribes lacked their own attorneys to represent them in protecting their water rights. Even when government attorneys in water litigation represented the tribes this representation was occasionally compromised.

For instance, when the federal Bureau of Reclamation decided to build an irrigation project for the benefit of non-Indian farmers using water from the

Truckee River in Nevada it was necessary to adjudicate the water rights of everyone on the river in order ensure the government had necessary water rights for the project. The Pyramid Lake Paiute Tribe's reservation was at the end of the river surrounding a large lake where the tribe traditionally fished to derive its subsistence. In the adjudication of water for the federal irrigation project, the government stepped in to represent the interests of the tribe as well as the Bureau of Reclamation. The government attorneys claimed irrigation water for the tribe's lands surrounding the lake but failed to claim any water rights for the tribe to sustain the fishery in the lake. As a result of the water project, the lake level dropped drastically, destroying the fishery and the livelihoods of Indian fishermen, and driving two species of fish to the brink of extinction. The Pyramid Lake case is one of many in which government conflicts of interest interfered with tribal water rights.

Non-Indian water development planned and paid for by the federal government often conflicts and competes with Indian water rights. This is ironic considering the well-established legal principle in American Indian law that the government is charged with a trust responsibility to act for the benefit of Indian tribes. The National Water Commission found in its 1973 report that:

> Following *Winters*, ... the United States was pursuing a policy of encouraging the settlement of the West and the creation of family-sized farms on its arid lands. In retrospect, it can be seen that this policy was pursued with little or no regard for Indian water rights and the *Winters* doctrine. With the encouragement, or at least the cooperation, of the Secretary of the Interior – the very office entrusted with the protection of all Indian rights – many large irrigation projects were constructed on streams that flowed through or bordered Indian Reservations, sometimes above and more often below the Reservations. With few exceptions the projects were planned and built by the Federal Government without any attempt to define, let alone protect, prior rights that Indian tribes might have had in the waters used for the projects. ... In the history of the United States Government's treatment of Indian tribes, its failure to protect Indian water rights for use on the Reservations it set aside for them is one of the sorrier chapters. (National Water Commission, 1973, pp. 474–5.)

Many decades after the Supreme Court first articulated the reserved water rights doctrine, Indian water rights finally gained considerable attention in 1963 when the Court issued its landmark opinion in *Arizona* v. *California*.[4] The case involved an allocation of the Colorado River's flow among three of the states that touch the river. The United States, which was involved in the case because the river also crosses extensive Indian and federal lands, claimed reserved rights along the river for five tribes. The Supreme Court awarded those tribes 900 000 acre-feet of water per year – enough for a city of several million – an amount determined by calculating how much water would be required to irrigate all of the 'practicably irrigable acreage' on the reservations. This ruling sent a strong message to water users all over the West that Indian claims could translate to formidable amounts of water. Although the reserved rights doctrine had been idle, it was far from dead.

Prior appropriation was, and still is, the prevailing method for allocating water in the American West. As discussed in Chapter 3 and elsewhere, this doctrine has been altered in various ways and embellished with rules aimed at satisfying important public purposes. Nevertheless, most of the West's water was long ago allocated to the earliest users of water. The most valuable rights are the oldest because, in times of shortage, the holders of those rights can insist on delivery of the full quantity of water to which they are entitled. Accordingly, when senior users assert their rights, the most junior users often must curtail their water use. The Supreme Court created the reserved water rights doctrine to fit into the priority system; with a tribe's priority date established by the date its reservation was established. Because most reservations were established more than one hundred years ago, the accompanying water rights are usually quite senior.

If a precise priority date is determined for a tribe's water right, other users can calculate which non-Indian users potentially must cut back their water use in order to allow water to flow to the reservation. But the scope of the right – and thus its impacts on other water users – remains uncertain until the quantity of water to which the tribe is entitled is decided. By contrast, non-Indians' water rights can be quantified because they exist under the prior appropriation doctrine only to the extent that water has been used. Reserved rights can exist without a history of actual use and then later be claimed by a tribe, potentially disrupting the uses of its non-Indian neighbors. The resulting lack of certainty can frustrate non-Indian water users when they seek to invest or borrow money based on assumptions about how much water is generally available to them.

One solution to this uncertainty is to quantify Indian reserved rights. This can be done judicially by asking a court to decide how much water is necessary to fulfill the purposes of a reservation. Where the purpose of setting up the reservation is to allow the Indians to pursue agriculture, the courts follow the formulation in *Arizona* v. *California*, based on the reservation's practicably irrigable acreage (PIA). In adopting the PIA formula, the Supreme Court opened the way for tribes to claim huge quantities of water, especially in arid areas where the amount of water needed to produce crops can be enormous. The Court expressly rejected the idea that tribes should get merely a 'fair share' of the water in a river or that rights should be determined based on reservation populations. The court said that rights were not to meet present needs, but to meet future needs and therefore should be set according to the reservation's full capacity to use water for irrigation.

A court seeking to determine how much land is irrigable and how much water is required for irrigation must examine evidence of soil type, structure, and depth, topography, salinity content, possible crops, and climate. As this information usually is based on expert studies in hydrology, soil science, engineering, and economics, trials can be long and expensive. Given the importance of scarce water, the process is often contentious.

Tribal water rights are determined by state courts

The United States has two separate court systems, state and federal. The individual court systems of the 50 states have local courts with general jurisdiction and appellate court systems. These state courts usually handle water rights matters arising within a particular state. The United States generally is not subject to the jurisdiction of state courts, and the principle of sovereign immunity provides that the United States cannot be sued without its consent. Thus, ordinarily state courts would not be able to adjudicate federally owned water rights. Similarly, Indian tribes are also considered sovereign governments with immunity from suit without their consent or the consent of the United States Congress.

Federal courts, with district courts sitting in every state and a separate system of appeals, have more limited jurisdiction than state courts. The primary task of federal district courts is adjudicating 'federal questions', including interpretation and application of federal laws and hearing cases where the parties are from different states (diversity of citizenship). These courts technically have jurisdiction to determine how much water a tribe would be entitled to use for a reservation established under a treaty or agreement with the United States. The United States Congress enacted the McCarran Amendment of 1952, however, stating that when a state court takes jurisdiction over the adjudication of all water rights in a river, the United States will waive its sovereign immunity to suit so that the state court can determine all federal water rights.[5] Congress recognized in the McCarran Amendment the importance to non-Indians of knowing the extent of water rights of others with whom they compete for water in times of shortage under the prior appropriation doctrine. The law applies to all water rights owned by the United States. Although the United States only holds title to Indian water rights in trust for the tribes, the Supreme Court has held that Congress intended to extend state jurisdiction over those rights whenever the rights to an entire river were being adjudicated.[6] This development caused great concern for tribes because they feared that state courts would be less equitable to them than federal courts. There is a history of tension between tribes and states. The Supreme Court described the situation of Indians relative to states in 1886: 'They owe no allegiance to the states, and receive from them no protection. Because of local ill feeling the people of the States where they are found are often their deadliest enemies'.[7]

After the Supreme Court made it clear that Indian water rights were subject to determination in state courts, many states initiated 'general stream adjudications' – legal proceedings involving sometimes tens of thousands of water rights claimants in an entire river basin. The cost and complexity of these proceedings have proved burdensome to everyone involved. Some of these adjudications have continued for over 20 years without nearing completion. Today, there are over 60 Indian water rights cases pending in state courts.

Although many Indians believed that state courts would not provide fair trials for their water rights claims, the results have been mixed. In most cases the tribes have been able to prevail on the United States as their trustee to furnish them with lawyers and expert witnesses. Alliances of government and tribal lawyers have presented cases competently to the courts. In some cases, the state courts have awarded tribes impressively large quantities of water. Yet the overall record is not reassuring to critics who say that relegating tribal rights to the mercies of state courts is bound to be unfair to Indians.

In the adjudication of the Big Horn River, for example, the Wyoming Supreme Court affirmed the right of tribes of the Wind River Reservation to some 400 000 acre-feet of water – most of the water in the river.[8] Undeniably, the amount of water, based on a lower court's determination of the amount of irrigable acreage on the reservation, was enormous. Yet the state supreme court rejected the tribes' claims for water to be used for mineral development, fisheries, wildlife, and aesthetics. The court also denied the tribes' attempt to extend their reserved water rights to groundwater. Many scholars and at least some other courts differ with each of these adverse holdings. Whether or not the state court erred in defining the scope of the tribes' reserved water rights, it awarded them enough water to overshadow the impacts of those parts of the decision. The state challenged the decision in the United States Supreme Court but the state court decision was upheld, although barely; the Justices on the Supreme Court were divided by a vote of four to four.

The *Big Horn* case is the only state court adjudication of Indian water rights that has proceeded through final judgment and appeal to the Supreme Court. But other state courts have handed down rulings in general stream adjudications, some favorable and some unfavorable to Indian tribes. In Arizona, the state supreme court has held that the treatment of groundwater under state law as a resource that is allocated and managed under a regime entirely separate from surface water could not affect any rights the tribes had to groundwater under the reserved rights doctrine because those rights are a matter of federal law.[9] In Idaho, however, the state courts have rejected tribal claims to exemption from the state adjudication process under the McCarran Amendment.[10]

Negotiated settlement of Indian water rights claims as an alternative to litigation

The results in state court adjudications of Indian water rights vary, but all are terribly costly and take years to conclude. Indeed, it is rare for these cases to reach a final resolution in the courts; ultimately, nearly all have required multiparty negotiations followed by congressional action. The combination of the unpredictability and burdens of litigation have induced all parties to pursue negotiation of some or all issues, although the dispute is typically joined in the framework of adjudication.

Since the 1980s there have been about 18 negotiated settlements of Indian water rights. Settlement negotiations usually commence after a tribe or the United States becomes involved in litigation with a state and the associated non-Indian water users. Sometimes this involvement is part of a general stream adjudication started by a state under the McCarran Amendment. Settlement negotiations also can follow litigation in federal courts brought by the tribe or the United States. In a few cases settlement negotiations have begun without litigation having been initiated.

Negotiations can provide an opportunity for all interested parties to participate. The process is most useful when there are factual disagreements based on technical data. Rather than dwell on these conflicts, the parties seek to craft a solution that will satisfy at least some of their respective needs. Instead of producing an all-or-nothing judgment that is a clear-cut victory for one side, negotiations seek ways to provide recognition for tribal water rights without jeopardizing existing water uses. Although the tribes may not receive the full quantities of water originally or potentially claimed, they often get money – mostly from the federal government – to enable them to build facilities to put their quantified water rights to use.

The 'lubrication' provided by federal funding has been a key element in most Indian water rights settlements. It has allowed tribes to secure not only paper water rights but also 'wet water' delivered through irrigation systems and pipelines for domestic supplies. At the same time, non-Indians have gained assurance that they can continue using water rights that are junior to tribal water rights. Sometimes federal or state funding is also assured for projects that benefit non-Indian water users. Federal funding is usually part of a settlement package, and therefore the agreement reached by the various parties in negotiation must be approved and monies appropriated by Congress. Thus, settlements are almost always accompanied by federal and sometimes state legislation.

One of the first settlements involved the Ak-Chin Indian Community in Arizona. The tribe had substantial water claims against non-Indian users, which took decades of negotiation and three acts of Congress to resolve. The tribe first agreed with the Secretary of the Interior to forgo its claims in exchange for 85 000 acre-feet of irrigation water provided by a federal well-field water project. Pumping the well water on the Ak-Chin reservation, however, would deplete the groundwater under the Papago Indian reservation. In order to avoid this problem, the Department of Interior renegotiated a water contract with an irrigation district, which received its water from the federally financed Central Arizona Project, to require deliveries of the district's surplus Colorado River water to the tribe.[11]

Another example of negotiated settlement arose in Colorado. Two tribes had substantial claims on rivers that had long been used by non-Indian farmers and that were among the most feasible sources for water for growing communities in the region. After assessing the prospects of the water users in litigation, the state sought a negotiated settlement instead of litigation. It soon became clear that both the Indians and non-Indians in the area supported a major Bureau of Reclamation project that could, if funded by the federal government, provide additional water to the region and facilitate a resolution of the claims.

A settlement was proposed that involved quantification of the tribes' rights on several rivers, creation of multimillion dollar development funds for the tribes, and construction of the Animas-La Plata water project. Congress approved the settlement in 1988, omitting a provision that might have allowed the tribes to market their water to non-Indians out of state.[12] Before the government commenced construction it was determined that the project would harm endangered species of fish as well as causing other environmental problems. Moreover, the cost was estimated to be $700 million. Opponents argued not only that the project would be environmentally damaging and fiscally irresponsible but that the primary beneficiaries of the project would be non-Indians. Twelve years after the settlement was approved, the project still had not been funded and so Congress revisited the matter. By then, both the proponents and environmentalists had proposed several modifications to make the project more environmentally benign. When Congress approved a reconfigured Animas-La Plata project in 2000, it limited non-Indian benefits, substantially reduced the estimated cost, and declared that it could be operated without conflicting with the Endangered Species Act.

Each settlement is different because the legal, geographic, and economic situations of tribes vary, as do the political factors. The ability of a tribe and its neighbors in one state to achieve a settlement will vary with the relative power of the members of Congress that represent that state. The receptiveness of Congress to settlements will also vary depending on the economic health of the federal government at the time a settlement package is presented. Notwithstanding the inevitable differences among them, a review of the Indian water rights settlements to date shows several common characteristics:

- Federal investment in water or water facilities. By providing funds to build dams and delivery works, the settlement can ensure delivery of water to both Indians and non-Indians.
- Non-federal cost sharing. A typical condition of providing federal funds is that state or local governments bear a portion of the cost of the settlement.
- Creation of tribal trust funds. Cash funds are usually appropriated for the use of the tribes. Sometimes the money is to be used for water development and sometimes it is available generally for economic development.
- Limited off-reservation water marketing. For various reasons, tribes that are entitled to water rights often cannot or do not want to use all of their water on their reservations. Leasing water for use by non-Indians off the reservation can provide cash income that can help build the tribe's economic self-sufficiency while allowing non-Indians to use water they need for a period of years. Under the legal systems governing water in the West, rights to use water can be exchanged, leased, or transferred with few restrictions beyond protection of other water rights holders. Denying tribes the right to profit from allowing others to use their water rights seems inequitable. Most settlement packages allow the tribes to market their water but nearly all restrict Indian transfers.
- Deference to state law. Often settlements require that Indian water use be subject to state water law, at least when the water is used off the reservation. Where two or more states have entered into a compact allocating the use of a shared river an Indian water rights settlement that provides for the shared river to be the source of water used to satisfy tribal water rights claims usually provides that the compact will govern water use.
- Concern for efficiency, conservation, and the environment. Less pervasive among the settlements, but included in many of them, is a provision for improving the efficiency of water use and advancing environmental values.
- Benefits for non-Indians. Indian water settlements often provide benefits for non-Indians as a way of giving them political viability. At a minimum, non-Indians receive certainty that their established water uses can continue and if the United States agrees to build water facilities they may get access to water for new uses. In some cases, non-Indians have been able to obtain federal funding for projects that otherwise would have been politically impossible. They have succeeded in the context of Indian water rights settlements, however, by 'wrapping their projects in an Indian blanket'.

Current legal issues

Finality of determinations of rights   One of the goals of non-Indians in seeking quantification of Indian rights is to provide the certainty they need in order to make investments and to borrow money to build water projects and to develop their lands. This was surely a motive behind the enactment of the McCarran Amendment. Tribes can also benefit from knowing the extent of their rights as they try to attract investments in water facilities and otherwise to realize value from their unused water rights. Yet some tribes whose water rights have been adjudicated have suffered the consequences of inadequate legal representation. Even if mistakes are made, they cannot later return to court and ask for their water rights to be adjusted because that would disrupt non-Indian expectations. The likelihood that once water rights are determined the result will be immutable raises a serious concern for any tribe embarking on a quantification of its water rights.

In two cases where tribes had their rights fixed in the past and wanted to reopen cases to expand their rights, the Supreme Court refused to allow any change. In *Arizona* v. *California* (1963) five tribes along the Colorado River were represented in court by the United States Department of Justice.[13] Attorneys for the United States failed to claim water for all of the tribes' practicably irrigable acreage. Thus, the tribes' water rights were limited to the quantity needed for the irrigable lands identified by the government lawyers. Many years later the tribes were able to hire their own lawyers and experts and reopened the case. They proved that additional lands were irrigable and asked the Supreme Court to award a greater quantity of water. But the Supreme Court held that the original quantification could not be changed except where the determination of practicably irrigable acreage had been caused by an error in boundaries that was corrected in a later court decision.[14] The Supreme Court said there is such a 'strong interest in finality' in western water law that it would be unfair to the non-Indians who had relied on the original quantification to allow the tribes to increase their claims.

The story of the Pyramid Lake Paiute Tribe that was discussed earlier furnishes another example. Because the federal government claimed for the tribe only water rights sufficient to irrigate the lands ringing the lake, and claimed no water to maintain the Indians' fishery, a federal project was built that diverted nearly all of the water from the single stream that supplied water to the lake. Without water to sustain the fishery and the lake level, the lake shrunk and the fish started to die off.

Years later in 1972 the tribe, through its own attorneys, proved that the United States had failed to claim sufficient water rights due to its conflict of interest and got a lower court to order the government to take action consistent with its trust responsibility and stop diverting all the water to the reclamation project.[15] The United States then attempted to reopen the old case that had given the tribe inadequate water rights. But in 1983, the United States Supreme Court refused to expand the tribe's water rights claims citing the interest of the non-Indians in having certainty in their water rights.[16]

Given the outcomes in these two cases, it is imperative for tribes whose reserved water rights are being determined by a court to participate in asserting the full extent of those rights. This is difficult for tribes with limited financial resources. In recent times, however, the United States has provided funding to some tribes for lawyers and experts even when the United States government also represented the tribes as a trustee.

The specter of a final and unalterable judgment may influence tribes not to seek an adjudication of reserved water rights. In most cases, however, tribes have no choice about whether to adjudicate their rights once the state initiates a general stream adjudication because the McCarran Amendment allows the state to join the United States as a party in such cases, and if the United States does not claim all federal and Indian water rights, they could be lost. Although the tribe, as a sovereign government, remains immune from being sued, the rulings of the Supreme Court teach that if the tribe abstains from the litigation it does so at its peril.

Water marketing   One of the most controversial questions concerning Indian water rights is whether tribes can sell or lease their water to non-Indians outside their reservations. In many cases, the government decided that Indians should become farmers, and moved them to reservations for that purpose. Some tribes, however, do not have a cultural tradition that is based on agriculture, or are unable to produce a livelihood because they were put on reservations that are too small or that have poor lands for farming. This has led some tribes to consider allowing others to use their water off the reservation. As explained earlier, most of the negotiated settlements of Indian water rights provide for some off-reservation use of tribal water rights, although these off-reservation uses are typically restricted in location and scope.

On many reservations non-Indians control the best agricultural lands. The government's nineteenth-century allotment policy opened up the reservations to non-Indian settlement; today, non-Indians cultivate 69 per cent of all farmland and have 78 per cent of the irrigated acreage on reservation lands throughout the nation.[17] Moreover, the allotments that were issued to individual Indians more than a century ago have descended through inheritance to an unwieldy number of heirs. Therefore, the only practical way to put these lands to use is to lease them, usually to non-Indian farmers. A share of the tribe's reserved water rights attaches to allotted lands and a lease to non-Indians can include the right to use water.[18]

There is considerable debate about whether tribes should have the legal right to allow their water rights to be used outside their reservations. Opponents of Indian water marketing argue that the nature of the reserved right is to make reservation lands useful and this purpose is not fulfilled when water is used elsewhere. Proponents say that the ultimate purpose of the reservations was to provide a homeland where Indians could be self-sufficient, a goal that may be best achieved if tribes can enter the marketplace and realize the economic value of tribal resources.

Off-reservation Indian water marketing also could provide a way to continue and expand non-Indian uses. If tribes could enter into contracts allowing others to use their water it could give greater certainty to non-Indian users. Nevertheless, some non-Indians who have long benefited from using undeveloped Indian water without charge oppose the idea of allowing tribes this right.

The most substantial legal question about Indian water marketing is whether a tribe has the legal right to convey what is essentially a property right without congressional approval. One of the oldest rules of Indian law (dating back to an 1823 decision) is that tribes cannot transfer land or interests in land to non-Indians without the participation or approval of the United States government.[19] Any legal doubt on this point can be resolved by obtaining congressional consent. This consent was granted in several negotiated Indian water settlements that allowed water marketing. Action by Congress also moots the issue of whether there is a fundamental conflict between the Supreme Court's original rationale for reserved water rights and a tribe's use of them outside the reservation. In any event the legal restraint on alienation of Indian property is intended to protect Indian rights from encroachment by non-Indians or the states. This suggests that the primary concern in whether Indian water should be marketable is whether the tribes have been dealt with fairly. Presumably, congressional approval should focus on ensuring that the transfer is in the best interests of the tribe.

Some observers have proposed that Congress should authorize tribes to lease their water rights subject to the approval of the Secretary of the Interior just as tribes can now lease tribal lands with secretarial approval. One of the arguments offered in favor of Indian water leasing is that non-Indians may freely transfer their water and water rights so long as the rights of others are not harmed and, therefore, it is inequitable to deny tribes the same ability to market their water.

Tribal water codes and administration   As sovereigns over their members and territory, Indian tribes can legislate and regulate the use of water within their reservations. Their ability to do so has been frustrated, however, by political impediments to the federal government's approval of tribal water codes as required in many tribal constitutions, and by some recent decisions of the Supreme Court that limit the reach of tribal regulatory authority over non-Indians on reservations.

It is clear that a state has no jurisdiction to regulate Indian use of Indian water rights. This is part of a 150-year legal tradition of maintaining tribal jurisdiction over Indians and their property on reservations, free from state control. The harder question is under what circumstances non-Indians on an Indian reservation can be controlled by tribes and when they fall under state jurisdiction.

Generally, if non-Indians are on Indian land they, like Indians, are subject to tribal jurisdiction. But some recent court decisions have created doubts about whether tribes can regulate non-Indians, especially if they are on non-Indian-owned land within the reservation. One case decided in 1981 says that a tribe may have jurisdiction over a non-Indian on its reservation, even on the non-Indian's fee land, if the non-Indian's conduct would threaten or have a 'direct effect on the political integrity, the economic security, or the health and welfare of the tribe'.[20] The use of waterways on a reservation presumably would affect some or all of these interests. But in a 1984 case dealing specifically with the applicability of a tribal water code, a federal court of appeals held that the tribe lacked the requisite interest to regulate.[21] The court reached this decision because the stream in question originated outside the reservation, ran only a short way along the reservation boundary, then turned away and joined the Spokane River outside the reservation. The same court later upheld the application of a tribal water code to non-Indians using water on their land within the reservation where the stream was entirely on the reservation.[22] The court added, however, that the tribe could not control 'excess' water used by non-Indians – presumably water not subject to re-served water rights.

It would appear that tribes with comprehensive, well-developed codes and regulations governing waters on their reservation would be better able to demonstrate the need to regulate non-Indian water to further tribal interests. For instance, the United States Supreme Court in 1983 upheld the exclusive authority of the Mescalero Apache Tribe to regulate game and fish on its reservation, including hunting and fishing by non-Indians.[23] This case did not deal with regulation on non-Indian land, but the court did emphasize the importance to the tribe of having unified regulation of a resource like wildlife. Similarly, the political integrity of tribal government control of resources would depend on unified control of water resources.

Tribes attempting to enact legislation to regulate water resources on their reservations have lacked support from the United States Department of the Interior. Perhaps half of the tribal constitutions have provisions that require certain tribal legislation to be approved by the Secretary of the Interior before it will take effect. For 26 years the department has maintained a moratorium on secretarial approval of any tribal water codes that would extend to non-Indian water use. On two occasions the department has circulated draft regulations governing the approval of such codes, but they were met with a firestorm of opposition from western senators and congressmen. The federal government has departed from the moratorium in only a few cases to approve tribal codes as part of negotiated water settlements approved by Congress.

Not all tribes have a secretarial approval requirement for tribal codes, and those that do may have a means to remove the impediment. Although the Clinton administration did not lift the moratorium, it voiced sympathy for the tribal effort to regulate water resources. Secretary Bruce Babbitt said that if a tribe wanted to enact a water code and confronted a requirement for secretarial approval, all the tribe had to do was to amend its constitution to remove the requirement for secretarial approval of ordinances, and he would approve the amendment removing the approval requirement. Thus, the tribe could adopt a water code without the need for federal approval. It is not known whether this will remain the policy of the current Bush administration, but the moratorium remains. Thus, even if a code is adopted without the necessity of secretarial approval, the apparent policy of the Department of the Interior disfavoring tribal codes could be cited if code enforcement is challenged by a non-Indian and a court is called upon to examine the tribe's authority to enact the provision.

Notwithstanding the uncertain area of tribal water code enforcement over non-Indians within a reservation, many tribes have sophisticated codes. Some tribal water resources department staffs include well-trained professionals who do water resources planning and enforce water rights among those who share in the use of water on the reservation.

Use of rights for new purposes   Reserved water rights can be quantified for any purpose for which the federal government established an Indian reservation. As described earlier, the most common purpose for creating reservations is to enable the Indians to pursue agriculture, but reserved rights can arise from other purposes. For example, a court of appeals in 1983 found that a treaty provision guaranteeing the Klamath Tribe the exclusive right to hunt, fish, and gather on its reservation reflected the primary purpose behind the creation of the reservation.[24] Other parts of the treaty mentioned agriculture and the court found that encouraging the Indians to take up farming was a second essential purpose of the reservation. Although state law did not allow water rights for fishing and hunting, the court held that the Indians had such a right which could be enforced to prevent non-Indians from depleting streams below levels that were required to maintain streamflows for fish and game.

A more difficult question arises when a tribe later wants to use water for purposes other than those for which its reserved water rights were quantified. For instance, if rights were quantified for agricultural uses, can a tribe use the water for industrial purposes, or for a fishery, or even to irrigate a golf course? Sixteen years after the Supreme Court had decided the reserved water rights of tribes on the Colorado River in *Arizona* v. *California* (1963), it accepted a report of a Special Master recommending that the tribes' use of water not be limited to the uses that were the basis of the original quantification.[25] This provides some indication that the Court might approve such changed uses in future cases.

Not all courts agree with the inference from *Arizona* v. *California* that tribes can use their water rights quantified for agriculture for other purposes. In the *Big Horn* adjudication, the court quantified the tribes' reserved rights based on irrigable acreage.[26] However, the Wind River Tribes decided to use a portion of these rights to restore streamflows within the reservation and build up the fishery. They recognized an opportunity to recover the natural ecosystem and to reap economic benefits from tourism and recreational uses by attracting anglers. Non-Indian water users on the reservation who would have had to leave water in the stream instead of diverting it for irrigation objected. The state supreme court in 1992 rejected the tribes' attempt to use water for instream flows, saying that any change in use would have to be in accordance with Wyoming state law, which does not recognize such instream uses as 'beneficial'.[27] The United States Supreme Court did not review the decision.

If a tribe decides to use its water rights for purposes that were not the basis for the quantification of reserved rights, this new use must be approved under state law. All western states apply the so-called 'no injury' rule to changes in use under the prior appropriation system. When non-Indians change their established uses they must show that no other water users are hurt by the change. But the rule does not fit well the situation of tribes whose rights under the reserved rights doctrine may not have actually been put to use. Most tribes have not had their rights quantified until recently and whether quantified or not, it has been difficult for the tribes to use water because they have been unable to raise the capital needed to develop water. On the Wind River Reservation, for instance, the federal government financed an irrigation system that served mostly non-Indians and the Indians have had little access to the system. Now the courts have held that the tribes have senior rights to water in the Big Horn River. Commencing any Indian use would displace hundreds of non-Indian water users who have been using water for years, thereby causing injury. The possible injury to established users would be worse if the tribes sought to restore streamflows to the river within the reservation but downstream of the irrigation project. Yet, to deny a tribe the ability to change to new uses would diminish the value of the water right.

The equities of established non-Indian water users nevertheless deserve consideration. The non-Indian irrigators are neighbors and they are not responsible for the way the system from which they benefit was developed and for the fact that it has operated to the detriment of the Indians. The government created the system and the non-Indians inherited it. They reasonably expected that the present conditions would continue. On the other hand, non-Indians have been using Indian water to build their own wealth. Under these circumstances, it seems inappropriate to apply the no injury rule mechanically. A rigid approach could halt tribal progress and further delay the long-denied tribal benefits from use of reserved water rights.

Walker and Williams (1991) propose that tribes like those on the Wind River Reservation exercise their authority to administer and regulate water rights and in doing so take control over the 'change of use' question. They can adopt criteria for 'sensible water use policies for all reservation citizens' non-Indian as well as Indian (Walker and Williams, 1991, p. 10). Some non-Indians have relied on state permits to use water diverted on the reservation that are over 80 years old. Walker and Williams (1991, p. 9) urge that tribes 'balance the complex interests of these non-Indians against … [t]he unique historical circumstances of water development on Indian reservations [that] may well compel compromise'. They suggest that one such compromise would be for tribes to adopt a public interest standard for tribal reservation water administration and apply it in a way that considers, along with other equities, the injury to juniors of changing the use of reserved water rights.

Critique and Conclusions

The doctrine of Indian reserved water rights is, in theory, a potent force for tribes. The combination of water rights recognized in the law of the dominant society, a land base where they can be exercised, and governmental powers to regulate how water is used is potentially very valuable. Yet application of doctrine has not justified the worries of non-Indians. And, accordingly, it has not fulfilled the promise it seemed to hold for Indian people in the United States. Only a handful of tribes – fewer than 30 – have finally determined the extent of their rights. Of those, only a few have put a significant portion of their water rights to use. Consequently, non-Indians have not been affected adversely by Indian water use. As Richard Collins (1985, p. 482) wrote:

> [T]his situation has generated powerful political and financial forces that oppose Indian development, of which there has been very little. There have been extravagant claims of the threat posed by Indian water claims, but actual conflict has been

almost entirely a war of words, paper, and lawyers. Indian calls are not shutting anyone's headgates.

The challenge to tribal leaders and their lawyers is to give Indian water rights doctrine potency in practice.

A major problem is that the processes for adjudication or negotiation to determine reserved water rights are expensive and arduous. The results are also uneven, depending as they do on the fortuity of how much political power a particular state's congressional delegation wields and the timing relative to the nation's economic well-being. Once tribal rights are quantified they often remain unused because of a shortage of capital, restrictions on marketing, and limits on changes of use.

In order to ensure that reservation water is used wisely, the tribes must also be able to exercise comprehensive control over it. With large blocks of land on many reservations owned by non-Indians, comprehensive control requires asserting jurisdiction over non-Indian users within the reservation, an effort that may confront strenuous legal opposition.

Achieving justice and equity for Indians, then, depends not only on having an ample legal foundation but also on having fair and reasonable means to develop, use, and regulate water resources. This requires political support that is difficult to achieve given the relatively greater influence of non-Indians. Moreover, there is a cultural divide that can make it difficult for non-Indian decision makers to appreciate or accept the legitimacy of Indian water claims.

2. LESSONS FROM AROUND THE WORLD

Indigenous people in nearly all countries have seen their traditional water sources dominated by non-native societies to produce economic benefits. Water law systems have been created that foster these uses and that are generally incompatible with traditional uses. Perpetuation of activities by indigenous people that depend on water is also frustrated by diminishment of the lands over which they have control. Although, as part 1 of this chapter explains, tribes in the United States have not fully realized the potential benefits of ostensibly strong water rights recognized in law, they do have distinct territories over which they can and do exercise a degree of sovereignty. They also have secured at least the nominal recognition of valuable water rights.

The experiences of certain other countries offer some ideas that should be considered to enhance the value of tribal water rights in the United States. On the other hand, tribal peoples in most countries lack the right to a defined land area and the recognized power to govern that United States tribes have. This may suggest that the United States system, despite its inadequacies and inequities, may have some advantages to the extent that these tribal rights and powers can be used to protect and enable use of water adequate for their present and future needs.

Australia

In Australia, indigenous peoples' rights received little legal recognition until the last decade. A 1992 decision of the Australian High Court in *Mabo* v. *Queensland*[28] recognized the existence of native title in land under the common law. It is now clear that Aborigines have interests in lands and waters held by the Crown. These rights include certain traditional uses such as hunting, fishing, and gathering. The notion of native title is a new concept and it has not yet been developed to the point that one can describe with certainty the nature of indigenous water rights.

Rights under native title can be lost when native people lose their connection with the land or by extinguishment when the government grants the land to others. After 1975, such extinguishments (and the resulting title in non-indigenous people) were subject to legal challenge under the Racial Discrimination Act passed that year, but the Native Title Act of 1993, among other things, shut off those arguments by validating past and future conveyances.[29]

The Native Title Act confirms that native title can be claimed in waters as well as lands. David Farrier (2002) argues that this opens the possibility that Aborigines can claim rights to use waters for everything from irrigation to fishing to cultural and spiritual uses and therefore that claims could be made to extract water for some uses and to maintain instream flows for others. Yet the courts have not resolved all of the issues necessary to sustain such claims.

It is likely that indigenous communities will be able to assert successfully their rights to traditional uses of water. In 2000, the Federal Court confirmed an agreement between parties which gave native title holders a right to take water for the purposes of satisfying their personal, domestic, social, cultural, religious, spiritual, or non-commercial communal needs, including the observance of traditional laws and customs.[30] However, a claim to water for irrigation might not fare as well. There is a common assumption that historically, traditional practices of Aboriginal communities did not include irrigation. In other contexts, however, the courts have said that they will not limit the rights under native title strictly to the customary uses that prevailed in ancient times but will allow, at least, 'the maintenance of the ways of the past in changed circumstances'.[31]

A serious problem with native rights to water, even for demonstrably traditional uses, is that they are subordinate to all of the rights that have been granted to others by the state under water legislation.[32] Furthermore, the very enactment of water legislation may pose problems for indigenous peoples seeking to assert rights to water. The legislation of the various states vests in the state the exclusive right to the use, flow and control of waters. In *Western Australia* v. *Ward* (2000) the Full Court held that such a provision did not extinguish native title rights in the course of abolishing all existing water rights and replacing them with statutory rights.[33] But the majority held that the legislation did operate to destroy any *exclusive* rights to control the use and enjoyment of water that the native title claimants could otherwise show. In Western Australia this leaves natives with non-exclusive rights to take and use water for personal, domestic, and non-commercial communal purposes.

To the extent that the enactment of water legislation itself did not extinguish water rights under native title, it could be argued that state issuance of water licenses effects an extinguishment. The *Ward* decision said that this would be possible if the intention to do so was 'clear and plain' – a criterion usually applied to a legislative act itself and not to executive acts implementing the act. Farrier (2002) believes that the highly discretionary nature of issuing water licenses makes it unlikely that clear and plain intent could be imputed to the legislature by such executive actions. In any event, he points out that the limitation on access in the Western Australia legislation involved in *Ward* might limit the exclusivity of a claim under native title, but should not effect a complete extinguishment.

Several states have recently enacted water management legislation but they vary in their impact on Aboriginal peoples (Farrier, 2002). In New South Wales legislation there are specific references to indigenous interests in water. A stated purpose of the act is to foster 'benefits to the Aboriginal people in relation to their spiritual, social, customary and economic use of land and water' resulting from the sustainable and efficient use of water. Despite this broad language, the likelihood of indigenous communities getting rights to water for commercial purposes or for instream flows may be illusory because the water courses are fully committed to non-indigenous uses. The New South Wales legislation does state that water management committees must include Aboriginal representatives, ensuring them some level of participation. But in South Australian legislation the interests of Aboriginal peoples are ignored. There is not even mention of including indigenous communities in water planning or watershed management boards.

Overall, the water claims of indigenous peoples are rather tenuous under the laws of Australia. The most substantial hope is that they can obtain licenses under the laws of specific states. It seems that a conflict inexorably will be confronted, given the concept in Australian indigenous law that proscribes future (post-1993) extinguishment of native title rights as racially discriminatory action. To avoid discriminatory results the government will have to reallocate water from non-indigenous people and that will require compensation if their rights are considered vested.

South Africa

> We cannot accept that the white men and their machines transform the river into a market currency for fast profit. So, human beings with ecological consciousness must create mechanisms to stop predator humans from continuing with this craziness. Water is a common asset.
>
> Marcos Terena, Terena Tribe, Coordinator General of
> Indigenous Rights, Mato Grosso, Brazil

Can the 'public interest' in water provide adequate protection for indigenous water rights? If indigenous peoples are to have no specifically recognized rights under the domestic law of a nation, they may be relatively better off if there are strong, concrete public protections for all citizens in the water law. Such laws elevate ecological concerns as well as basic rights to put water to essential human uses over the property concepts and utilitarian values that are reflected in most water rights systems. Perhaps the most impressive public rights language of any country is found in South Africa. We discuss it here briefly, not because it specifically addresses indigenous rights, but because it keeps open the possibility that indigenous peoples in largely colonized areas where most water is already dedicated to economic uses can assert the public interest to protect water uses for subsistence, ceremonial and community needs. (The South African water law is discussed in more detail in Chapter 3.)

In the post-apartheid era, South Africa substantially revamped its water law. The former system produced harsh and racially discriminatory results for the non-white majority. Most people were without access to safe drinking water. The cornerstone of water reform is found in the new South African Constitution. The Bill of Rights guarantees every South African the right to have access to sufficient water. The state must take reasonable legislative action and other measures, within its available resources, to achieve the progressive realization of this right. Furthermore, everyone is guaranteed a right to an environment that is not harmful to human health or well-being and to have the environment protected, for the benefit of present and future generations.

The new water law extends the idea of the 'public trust' that has had judicial recognition in California and, to a lesser extent, in a few other states (as explained in Chapter 3). In the Mono Lake litigation, the California Supreme Court announced the ideal of an 'integrated system of preserving the continuing sovereign power of the state to protect public uses, a power which precludes anyone from acquiring a vested right to harm the public trust, and imposes a continuing duty on the state to take such uses into account in allocating water resources'.[34] But unlike any jurisdiction in the United States, this doctrinal ideal is given legislative substance in South Africa. The National Water Act defines national government, acting through the Minister of Water Affairs and Forestry ('the Minister'), as the *'public trustee'* of the nation's water resources.[35] The minister is ultimately responsible for ensuring that water is protected, used, developed, conserved, managed and controlled in a sustainable and equitable manner, for the benefit of all persons and in accordance with its constitutional mandate. Failure to fulfill the South African public trust obligations makes any resultant decision subject to administrative review by either a special Water Tribunal or the courts.

One feature, and a notable innovation, of the South African national Water Law is the 'Reserve', which is an unallocated quantity of water that is not subject to competition with other water demands. The Reserve is made up of 'that quantity and quality of water required to (a) satisfy basic human needs for all people who are, or may be supplied from a relevant water resource; and (b) protect aquatic ecosystems in order to ensure ecologically sustainable water development and use'.[36] Although the purpose is to serve broad public interests, one can perceive the benefits to indigenous peoples who would otherwise be left to compete in a system where they lack economic or political power. Such a law would not, however, allow for *economic* uses by native peoples except to the extent of allowing equal access to water sources. Thus, it would not be as complete protection as the present "reserved rights" doctrine theoretically is in the United States.

International Law

There are several international agreements and documents that express norms for treatment of indigenous rights in water allocation. In countries that subscribe to them, these instruments may provide protections not available in domestic law. For instance, indigenous people may be able to assert claims for denial of access to water, exclusion from the decision-making process, and disregard for their traditional systems for allocating and using water. International conventions of potential applicability include the American Convention on Human Rights; the International Covenant on Economic, Social and Cultural Rights; the International Convention on the Elimination of All Forms of Racial Discrimination; the International Covenant on Civil and Political Rights; the International Labour Organization Convention No. 169; and the United Nations Convention on Biological Diversity. Furthermore, the United Nations Draft Declaration on the Rights of Indigenous Peoples would extend an array of protections to indigenous peoples, including the right to their culture, land, and resources.

Courts in some countries have upheld the application of international law to protect the interests of indigenous peoples. For instance, Australia enacted a statute prohibiting dam construction planned by the State of Tasmania in an area listed as a World Cultural and National Heritage site under the World Heritage Convention.[37] The High Court upheld the legislation and said that the Convention imposed an obligation on Australia to take appropriate measures for the preservation of the site.

Article 27 of the International Covenant on Civil and Political Rights (ICCPR) guarantees indigenous people the right to enjoy their indigenous cultures. When the Japanese government constructed a dam in an area where indigenous Ainu ceremonial places had been, two Ainu sued. The Sapporo District Court held that the construction impaired the Ainu culture and the government violated a right to enjoy their indigenous culture under the ICCPR. Morihiro Ichikawa (2002) argues that it is possible further to construe Article 27 of the ICCPR as securing Ainu fishing rights and water rights as a part of their culture. Ichikawa believes that similar applications of international laws could assist indigenous peoples resisting dam construction. He cites the example of the Ainu where a paved road to be constructed in wild areas within Daisestuzan National Park in 1995 would have breached the Japanese government's obligation 'not to destroy biodiversity' under the Convention on Biological Diversity.

Besides formal agreements, a variety of norms that would apply to indigenous water rights has been accepted by the nations of the world. For instance, 'Agenda 21' adopted at the 1992 United Nations Conference on Environment and Development, provides a set of standards for countries in their use and conservation of natural resources. One standard requires the full participation of the public, especially water user groups and indigenous people and their communities.[38]

The Stockholm Declaration, adopted at the 1972 United Nations Conference on the Human Environment, also implies that indigenous peoples have a right to an equitable share of a nation's waters. Principles 5 and 21 express a common conviction that while 'states have … the sovereign right to exploit their own resources pursuant to their own environmental policies', they '… must be employed in such a way … to ensure that benefits from such employment are shared by all mankind'.[39]

A right to water for indigenous peoples seems to be included in the Vienna Declaration, adopted by the 1993 United Nations World Conference on Human Rights. Principle 20 of the declaration 'recognizes the inherent dignity and the unique contribution of indigenous peoples … and strongly reaffirms the commitment of the international community to their economic, social and cultural well-being …'.[40] Given the link between water rights and the well-being of indigenous peoples, denial of access to sources to meet basic water needs would seem to be contrary to this declaration.

Furthermore, the Declaration of Principles of Indigenous Rights adopted by the Fourth General Assembly of the World Council of Indigenous Peoples calls for legal recognition of indigenous water rights. This declaration states that the 'customs and usages of the indigenous peoples must be respected by the nation-states and recognized as a legitimate source of rights' and that indigenous peoples have 'inalienable rights over their traditional lands and resources'.[41] This declaration offers strong support for indigenous water rights based on traditional use.

Finally, because Article 38 of the Statute of the International Court of Justice states that custom is one of the primary sources of international law, it is important to note that several countries recognize indigenous water rights either through domestic or international law. John Alan Cohan (2001, p. 154) has described the current state of customary international law, asserting that 'the international community now regards indigenous peoples as having environmental rights that rise to the status of international norms' and that '... because indigenous peoples' way of life and very existence depends on their relationship with the land, their human rights are inextricable from environmental rights... .' He argues that these environmental rights include 'the right of indigenous peoples to control their land and other natural resources ... to maintain their traditional way of life'.

The utility of international agreements and norms for indigenous peoples varies among countries based on the degree to which their courts accept international law as a rule of decision. This acceptance is especially limited in United States courts. First, the United States has not ratified many of these agreements. Moreover, United States courts rarely apply international law even when the United States has subscribed to particular instruments. Curtis Bradley (2000, p. 246) explains that '... although there are a variety of specific reasons why [US] courts have rejected [international law] claims ... it appears that, in general, courts are more resistant to allowing international human rights claims when they are brought against domestic defendants.' Furthermore, access to international tribunals is restricted. Professor Williams (1990, pp. 695–6) concludes that 'the International Court of Justice and many other more effective and high-profile forums of international law are available only to states – a term which under present conceptions of international law, does not include indigenous peoples.' Nevertheless, the use of international law may be the most viable basis for advancing indigenous water rights in countries where domestic law is lacking and international law is accepted as the basis for decision making.

3.   ANALYSIS AND CONCLUSIONS

Generally speaking, the values to be secured by water law affecting indigenous water rights are: access to healthy sources of water, protection of flowing water needed to sustain ecosystems fundamental to indigenous economies and cultures, equitable treatment, and sufficient local autonomy to give meaning to customs and traditions concerning the use and allocation of water. Because indigenous peoples are almost invariably subject to the legal system of another culture that has gained its sovereignty through colonization, conquest, or otherwise without the consent of the native peoples, it is difficult to achieve legal protection for water uses that allows all of those values to flourish.

We have not attempted to make a comprehensive survey of water laws throughout the world and how they may affect the values important to indigenous peoples. But we can offer the observation based on our limited review that the US system of reserved water rights has some remarkable strengths. To be sure, our enthusiasm for the system is tempered by the reality that while it creates substantial legal rights on paper it lacks tangible success in securing 'wet water' for use by a significant number of tribes. Whether the doctrine in the United States will prove to be an effective tool for economic and cultural survival depends on the will of the federal government in enforcing the law and the integrity of the court system in upholding venerable principles. Even if tribes do succeed to the detriment of non-Indian water users, a political backlash could force Congress to abrogate Indian water rights as they can do under established principles of federal Indian law. Therefore, we caution that even the respectable but limited success of the United States system is precarious.

General acceptance of Indian water rights, however, may increase as notions of public rights and the public interest are embraced by states. Most of the values furthered by enforcement of tribal rights are compatible with the goals of water law reform in the United States and elsewhere, if not the short-term economic goals of those countries. As discussed throughout this book, water law is evolving in many ways to respect and advance equitable distribution, ecologically sound use, and public participation, especially at the local level. This is the essence of the trend toward recognition of the 'public interest'. The application of Indian water rights should promote fairer allocation by providing water to aid economic development on reservations, help to ensure streamflows for healthy fish populations that support traditional pursuits and ceremonial uses, and allow for local tribal governance of reservation resources.

NOTES

1. *Winters v. United States*, 207 U.S. 564, 576 (1908).
2. *Winters v. United States*, 207 U.S. 564, 576 (1908).
3. *Arizona v. California*, 373 U.S. 546 (1963).
4. *Arizona v. California*, 373 U.S. 546 (1963).
5. 43 U.S.C.A. § 666.
6. *Colorado River Water Conservation District v. United States*, 424 U.S. 800, 810 (1976).
7. *United States v. Kagama*, 118 U.S. 375, 384 (1886).
8. *In re General Adjudication of All Rights to Use Water in the Big Horn River System*, 753 P.2d 76, affirmed, *Wyoming v. United States*, 492 U.S. 406 (1989).
9. *San Carlos Apache Tribe v. County of Maricopa*, 972 P.2d 179 (Ariz. 1999).
10. *In re Snake River Basin Water System*, 764 P.2d 78 (Idaho 1988).
11. Public Law No. 95–328, 42 Stat. 409 (1978); Public Law No. 98–530, 98 Stat. 2698 (1984); Public Law No. 102–497, 106 Stat. 3528 (1992).
12. The project was authorized in the Colorado River Basin Project Act of 1968 (P.L. 84–485; 70 Stat. 105) and later incorporated into the Colorado Ute Indian Water Rights Settlement Act of 1988 (P.L. 100–585; 102 Stat. 2973). Due largely to environmental and fiscal issues, the project did not move forward until modified further in the Colorado Ute Settlement Act Amendments of 2000 (enacted within the Consolidated Appropriations Act of 2001, P.L. 106–554, by reference from H.R. 5666).
13. 373 U.S. 546 (1963).
14. *Arizona v. California*, 460 U.S. 605 (1983).
15. *Pyramid Lake Paiute Tribe of Indians v. Morton*, 354 F.Supp. 252 (D.D.C. 1972).
16. *Nevada v. United States*, 463 U.S. 110 (1983).
17. *Water Law and Indigenous Rights (WALIR) Towards Recognition of Indigenous Water Rights and Management Rules in National Legislation*, Summary of Presentation at the Public Meeting (7 March 2002), Wageningen, The Netherlands.
18. *Skeem v. United States*, 273 Fed. 93 (9th Cir. 1921); 25 U.S.C. § 415.
19. *Johnson v. McIntosh*, 21 U.S. (8 Wheat.) 543 (1823); Non-Intercourse Act, 25 U.S.C.A. § 177.
20. *Montana v. United States*, 450 U.S. 544, 566 (1981).
21. *United States v. Anderson*, (736 F.2d 1358 (9th Cir. 1984).
22. *Holly v. Confederated Tribes & Bands of the Yakima Indian Nation*, 655 F. Supp. 557 (E.D. Wash. 1985), affirmed subnom, *Holly v. Totus*, 812 F.2d 714 (9th Cir. 1987), certiorari denied, 484 U.S. 823 (1987).
23. *New Mexico v. Mescalero Apache Tribe*, 462 U.S. 324 (1983).
24. *United States v. Adair* (753 D.2d 1394 (9th Cir. 1983), certiorari denied, *Oregon v. United States*, 467 U.S. 1252 (1984).
25. *Arizona v. California*, 439 U.S. 419 (1979).
26. 753 P.2d 76, 98 (Wyo. 1988), affirmed, *Wyoming v. United States*, 492 U.S. 406 (1989).
27. *In re General Adjudication of All Rights to Use Water in the Big Horn River System*, 835 P.2d 273 (Wyo. 1992).
28. 175 CLR 1 (1992).
29. Racial Discrimination Act, Austl. C. Acts, No. 52. (1975); Native Title Act of 1993, Act No. 110, Office of Legislative Drafting, Canberra.
30. *Mark Anderson on behalf of the Spinifex People v. State of Western Australia* (2000) FCA 1717.
31. *Members of the Yorta Yorta Aboriginal Community v State of Victoria* (2001) 180 ALR 655; (2001) FCA 45.
32. *Ngalpil v. State of Western Australia*, (2001) FCA 1140.
33. 170 ALR 159; (2000) FCA 191.
34. *National Audubon Society v. Superior Court*, 33 Cal.3d 419,189 Cal.Rptr. 346 (1983).
35. National Water Act, 20 Aug. 1998, No. 36, South Africa.
36. National Water Act, 20 Aug. 1998, No. 36, Section 1(xviii).

37. Water Management Act 2000 (New South Wales, § 55(2)(a).
38. UN Conference on Environment and Development, Agenda 21, Chapter 18.34(d).
39. Declaration of the United Nations Conference on the Human Environment, 1972, Principle 5, available at www.unhchr.ch/huridocda/huridoca.nsf/(Symbol)/A.CONF.157.23. En?OpenDocument
40. Declaration of the United Nations Conference on the Human Environment, 1972, Principle 20, available at www.unhchr.ch/huridocda/huridoca.nsf/(Symbol)/A.CONF.157.23. En/OpenDocument
41. The full declaration can be viewed online at www.cwis.org/fwdp/Resolutions/WCIP/wcip.txt

LITERATURE CITED

Bradley, Curtis A. (2000), 'Customary international law and private rights of action', *Chicago Journal of International Law*, **1**, 421.

Cohan, John Alan (2001), 'Environmental rights of indigenous peoples under the Alien Tort Claims Act, the public trust doctrine and corporate ethics, and environmental dispute resolution', *UCLA Journal of Environmental Law and Policy*, **20**, 133.

Collins, Richard B. (1985), 'The future discourse of the winters Doctrine', *University of Colorado Law Review*, **56**, 481.

Farrier, David (2002), 'Protecting environmental values in water resources in Australia and a note on indigenous rights to water in Australia', paper prepared for the Natural Resources Law Center, University of Colorado School of Law, Boulder, CO.

Getches, David H., Charles F. Wilkinson and Robert A. Williams, Jr (1998), *Cases and Materials on Federal Indian Law*, St. Paul, MN: West Publishing Co.

Ichikawa, Morihiro (2002), 'Protection of ecological and cultural values of watersheds under the Convention on Biological Diversity and the International Covenant on Civil and Political Rights', paper prepared for the Natural Resources Law Center, University of Colorado School of Law, Boulder, CO.

National Water Commission (1973), *Water Policies for the Future*, Port Washington, NY: Water Information Center, Inc.

Walker, Jana L., and Susan M. Williams (1991), 'Indian reserved water rights', *Natural Resource and the Environment*, **5**, 6.

Wiener, John (2002), 'Destroying (by not integrating) culture and environment: the legal implications of the common property movement and some notes on the ditches of Colorado', paper prepared for the Natural Resources Law Center, University of Colorado School of Law, Boulder, CO.

Williams, Robert A. Jr (1990), 'Encounters on the frontiers of international human rights law: redefining the terms of indigenous peoples' survival in the world', *Duke Law Journal*, **1**, 660.

5. Transboundary water conflicts and cooperation

Aaron T. Wolf

With assistance from Robert K. Hitchcock, University of Nebraska; Jeffrey Jacobs, National Research Council; Mikiyasu Nakayama, Tokyo University of Agriculture and Technology; Julie Trottier, Oxford University; and Douglas S. Kenney, University of Colorado

INTRODUCTION

Transboundary water disputes can be defined broadly as occurring whenever demand for water is shared by any sets of interests, be they political, economic, environmental, or legal. Conflicts over shared water resources occur at multiple scales, from sets of individual irrigators, to urban versus rural uses, to users located in different political jurisdictions – the traditional definition of transboundary. Transboundary waters share certain characteristics that make their management especially complicated, most notable of which is that these basins require a more complete appreciation of the political, cultural, and social aspects of water, and that the tendency is for regional politics to regularly exacerbate the already difficult task of understanding and managing complex natural systems.

International transboundary water issues are increasingly being viewed through the lens of security studies, which are guided by an appreciation of the mutually destabilizing forces of poverty and political instability. The process of poverty alleviation is often hampered in regions where human security is at risk. As a consequence, much of the thinking about the concept of 'environmental security' has moved beyond a presumed causal relationship between environmental stress and violent conflict to a broader notion of 'human security' – a more inclusive concept focusing on the intricate sets of relationships between environment and society.

Within this framework, water resources – including scarcity, distribution, and quality – have been named as the factor most likely to lead to intense political pressures, while threatening the processes of sustainable development and environmental protection. Water ignores political boundaries, evades institutional classification, and eludes legal generalizations. Worldwide, water demands are increasing, groundwater levels are dropping, water bodies are increasingly contaminated, and delivery and treatment infrastructure is aging.

From the Klamath to the Jordan, transboundary water issues are a priority at state, national, and international levels. Although full-fledged wars over water have not occurred in modern history, there is ample evidence showing that the lack of clean freshwater has been linked to poverty and has led to intense political instability, and that acute violence has occasionally been the result at this scale. While these disputes also occur at the subnational level, the human security issue in this case is more subtle and more pervasive. As water quality degrades – or quantity diminishes – over time, the effect on the stability of a region can be unsettling, nowhere more so than in basins which cross political boundaries.

There are 261 watersheds which cross the political boundaries of two or more countries. These international basins cover 45.3 per cent of the land surface of the earth, affect about 40 per cent of the world's population, and account for approximately 60 per cent of global river flow (Wolf et al, 2003). Disparities between riparian nations – whether in economic development, infrastructural capacity, or political orientation – add further complications to water resources development, institutions, and management. As a consequence, development, treaties, and institutions are regularly seen as, at best, inefficient, often ineffective, and, occasionally, as a new source of tensions themselves. Despite the tensions inherent in the international setting, riparians have shown tremendous creativity in approaching regional development, often through preventive diplomacy, and the creation of "baskets of benefits" which allow for positive-sum, integrative allocations of joint gains. Some of these approaches may be 'scalable', and relevant to the problems of the American West.

1.   TRANSBOUNDARY WATERS OF THE WEST

International Waters

There are two sets of international rivers in the American West: those shared between the United States and Canada, primarily the Columbia, and those shared between the United States and Mexico, especially the Colorado and the Rio Grande/Rio Bravo. Each is administered through different institutional structures – the International Joint Commission in the case of United States–Canada, and the International Boundary and Waters Commission for

United States–Mexico – and thus are described and assessed separately. A relatively new body of trinational law also exists in the region associated with the promotion of free trade.

United States/Canada waters

Canada and the United States share one of the longest boundaries in the world, at approximately 4000 miles. Industrial development in both countries, which in the humid eastern border region relied on water resources primarily for waste disposal, had led to decreasing water quality along their shared border to the point where, by the early years of the twentieth century, it was in the interest of both countries to seriously address the matter. Prior to 1905, only ad hoc commissions had been established to deal with issues relating to shared water resources as they arose. Both countries considered it within their interests to establish a more permanent body for the joint management of their shared water resources.

As Canada and the United States entered into negotiations to establish a permanent body, the tone was informed by the concerns of each nation. For the United States, the overriding issue was sovereignty. While it was interested in the practical necessity of an agreement to manage transboundary waters, it did not want to relinquish political independence in the process. This concern was expressed by the United States position that absolute territorial sovereignty be retained by each nation for the waters within its territory – tributaries should not be included in the Commission's authority. The new body might retain some of the ad hoc nature of prior bodies, so as not to acquire undue authority. Canada was interested in establishing an egalitarian relationship with the United States. This goal was hampered not only because of the relative size and level of development of the two nations at the time, but also because Canadian foreign policy was still the purview of the United Kingdom – negotiations had to be carried out between Ottawa, Washington, and London. Canada wanted a comprehensive agreement, which would include tributaries, and a Commission with greater authority than the bodies of the past.

The 'Treaty Relating to Boundary Waters between the United States and Canada', signed between the United Kingdom and the United States in 1909, reflects the interests of each negotiating body.[1] The treaty establishes the International Joint Commission (IJC) with six commissioners, three each appointed by the governments of Canada and the United States. Canada accepted United States sovereignty concerns to some extent – for example, tributary waters are excluded. The United States in turn accepted the arbitration function of the commission and allowed it greater authority than it would have liked. The treaty calls for open and free navigation along boundary waters, allowing Canadian transportation also on Lake Michigan, the only one of the Great Lakes not defined as boundary water. Although it allows each nation unilateral control over all of the waters within its territory, the treaty does provide for redress by anyone affected downstream. Furthermore, the commission has 'quasi-judicial' authority: any project that would affect the 'natural' flow of boundary waters has to be approved by both governments. Although the commission has the mandate to arbitrate agreements, it has never been called to do so. The commission also has investigative authority: it may have development projects submitted for approval, or be asked to investigate an issue by one or another of the governments. Commissioners act independently, not as representatives of their respective governments.

In 1944, the United States and Canada both asked the IJC to study the feasibility of cooperative development in the Columbia Basin, a process which lasted 20 years, until the signing of the Columbia River Treaty in 1961 and the subsequent establishment of coordinated river management rules (1961–64).[2] The focus of the treaty is a series of dams subsequently built for hydropower generation and flood control along the main stem and tributaries. The length of the negotiations reflect disagreements both within nations – notably in the United States between upstream states of Idaho and Montana, where the most inundation would have occurred, and downstream in Washington and Oregon where the bulk of the benefits would be realized – as well as between the United States and Canada. A budding environmental movement, concerned with loss of salmon runs, winter elk habitat, and the inundation of national parks, also played a role. Many of these concerns remain today.

The treaty stipulates: (1) the equal sharing of downstream benefits from hydropower and flood control in the United States that result from upstream storage in Canada; (2) the three storage sites in Canada, including the total volume for Treaty implementation (15.5 million acre-feet); (3) an option for the United States to build the Libby storage project; (4) the method, amount, and timing of United States payments to Canada; (5) the permissibility to transfer water from the Kootenay to the Columbia, including the timing and the maximum volumes to be transferred; (6) the option to transfer water out of the Columbia Drainage Basin; (7) the sequence of steps to be taken for conflict resolution if difficulties arise during treaty operations; and (8) the creation of new and/or designation of existing institutions to supervise and operate the treaty.

The United States Entity is composed of the Bonneville Power Administration (BPA) and the North Pacific Division, Corps of Engineers (COE), while the Canadian Entity is the British Columbia Hydro and Power Authority (BCH). The entities work through committees equally represented by members from each entity. The Operating Committee is instrumental in the planning and execution of treaty reservoir operations covered under the treaty.

While the treaty has been effective in managing water and power according to the priorities set during initial negotiations, many concerns of the day, as well as a host of new issues brought on by changing needs, growing populations, and increasing environmental awareness, remain.

United States/Mexico waters

The border region between the United States and Mexico has fostered its share of surface-water conflict, from the Colorado to the Rio Grande/Rio Bravo. It has also been a model for peaceful conflict resolution, notably through the work of the International Boundary and Water Commission (IBWC), the supralegal body established to manage shared water resources as a consequence of the 1944 United States–Mexico Water Treaty.[3]

The IBWC has its roots in the 1848 Treaty of Guadalupe Hidalgo, which established a temporary joint boundary commission to mark and map the new boundary between the two countries.[4] An 1889 convention established the International Boundary Commission, charging it with resolving '... differences or questions that may arise on that portion of the frontier between the United States of America and the United States of Mexico where the Rio Grande and the Colorado Rivers form the boundary line ...'.[5] The commission's status was permanently extended in 1900.

The 1944 treaty between the United States and Mexico firmly established the international character of waters on the border between the United States and Mexico. It specified in considerable detail the amount of water allocable to each country from the boundary rivers and their tributaries, with detailed delivery schedules and procedures for water accounting. Additionally, the treaty established the framework for construction of international storage reservoirs, diversion dams, and flood control works. This treaty also clearly established the role of the IBWC as the international organization the two countries would rely on in addressing these transboundary water issues.

The IBWC consists of a Mexican Section, headquartered in Ciudad Juarez, Chihuahua and a United States Section, headquartered just across the Rio Grande in El Paso, Texas, the midpoint along the international border. Each section is headed by an engineer commissioner appointed by the president of his country and operates under the guidance of each country's respective foreign affairs department.

The first water distribution treaty between the two countries, the Convention of 1 March 1906, established an agreed-upon amount of Rio Grande water allotted to Mexico at Ciudad Juarez, Chihuahua. This international agreement determined the national ownership of waters for the upper 145 kilometers of the Rio Grande's international segment. Decades later, in 1944, the national ownership for the remaining 1874 kilometers of the Rio Grande downstream to the Gulf of Mexico was established along with the authority to jointly construct impoundment and other engineering works for each country to make the greatest beneficial use of its apportioned waters.

Achieving these treaty allocations was a difficult process, marred by an incident responsible for adding the 'Harmon Doctrine' to the lexicon of international waters. Named for the United States Attorney General who suggested this stance in 1895 regarding a dispute with Mexico over the Rio Grande, the doctrine argues that a nation has absolute rights to water flowing through its territory (LeMarquand, 1993; McCaffrey, 1996).[6] Considering this doctrine was immediately rejected by Harmon's successor and later officially repudiated by the United States (McCaffrey 1996), was never implemented in any water treaty (with the rare exception of some internal tributaries of international waters), was not invoked as a source for judgment in any international water legal ruling, and was explicitly rejected by the international tribunal over the Lac Lanoux case in 1957, the Harmon Doctrine is wildly overemphasized as a principle of international law. Nevertheless, upstream nations, states, territories, and even individual landowners to this day regularly call on some variation of the Harmon Doctrine in the opening stages of negotiations.

Those treaty provisions related to the Colorado River and the practical effects of their implementation remain an ongoing source of discussion between the two countries. Over the past half century, various differences have arisen which required substantial attention from the IBWC in order to reach a satisfactory conclusion. The treaty provides a special annual allotment to Mexico and obligates the United States to provide that water under annual schedules provided by Mexico. There are provisions for times of excess flows and for times of shortages. In addition the treaty provides for works for the control of flood waters and for diversion structures by Mexico.

During the 1950s, the United States regularly made surplus declarations. However, as river conditions changed in the 1960s, the United States determined that no surplus existed. Mexico, having become accustomed to the surplus deliveries, expressed an interest in continuing to receive the larger deliveries. Mexico was also accustomed to receiving water with salinity adequate for their irrigation uses. The lower flows matter was complicated with the introduction from an irrigation district in Arizona of pumped saline drainage, which nearly tripled the salinity in waters delivered to Mexico. The salinity problem was dealt with through five-year arrangements of the IBWC supported by expertise from United States and Mexican federal agencies. The problem arose again in 1972, leading to a special Presidential task force and the adoption of 'Minute 242' of the international treaty in 1973 and enactment of the Colorado River Basin Salinity Control Act in 1974 (Holburt, 1975). Collectively, these laws provide the current framework for salinity management along the border.

In the 1980s, questions arose over surplus waters and their impacts in Mexico, a matter that was dealt with through a new technical information exchange program of the IBWC. Similarly, questions arose in the 1990s over silt deposition and flood water conveyance and salinity peaks in the waters delivered to Mexico. The IBWC turned its information exchange program into proactive international task forces to deal with the salinity problem, the immediate silt problem, and the longer-term conveyance questions. More recently, the IBWC has extended its information development task forces to a fourth group dealing with the Colorado River Delta.

Another more recent complication is the difficulties encountered in managing shared groundwater, which can pale in comparison to trying to allocate surface water resources. Each aquifer system is generally so poorly understood that years of study may be necessary before one even knows what the bargaining parameters are. Mumme (1988) has identified 23 sites in contention in six different hydrogeologic regions along the 3300 kilometers of shared boundary. While the 1944 treaty mentions the importance of resolving the allocations of groundwater between the two states, it does not do so. In fact, shared surface-water resources were the sole focus of the IBWC until the early 1960s, when a United States irrigation district began draining saline groundwater into the Colorado River and deducting the quantity of saline water from Mexico's share of freshwater. In response, Mexico began a 'crash program' of groundwater development in the border region to make up the losses. These tensions have resulted in renewed interest in resolving these topics.

An interesting aspect of the various IBWC agreements is the way in which binational projects are funded. In the case of the system to deliver Colorado River water to Mexico, the treaty required Mexico to pay for some works in the United States to protect United States interests from flooding. In addressing salinity issues, the United States agreed to pay for works in Mexico. Flexibility in allocating costs based on the benefits accrued to each country and the cost each country would incur if a project were domestic rather than binational are among the factors considered by the IBWC in determining a fair and equitable cost distribution that may or may not result in a 50–50 cost share. This has allowed the IBWC to deal with significant questions in a cooperative manner.

Trinational arrangements

As North America increasingly embraces free trade, a variety of trinational agreements and organizations are emerging that, theoretically, can have some influence on transboundary water resources. Of particular note is the North American Free Trade Agreement (NAFTA), a regional extension of the General Agreement on Tariffs and Trade established in 1947. While considerable ambiguities exist, these agreements likely do not provide for private trade in bulk water among the three nations, and thus respect existing treaty arrangements. Nonetheless, the influence of expanded trade on water resources is acknowledged, as evidenced by the establishment in 1994 of the trinational Commission on Environmental Cooperation, charged with finding long-term solutions to border environmental problems. The scope of this organization does not typically extend beyond border pollution issues, as larger-scale natural resource management concerns were 'formally and deliberately' omitted from its mandate (Mumme, 1999, p. 166).

Interstate Rivers

In addition to the rivers extending into Canada and Mexico, the United States is also home to many interstate rivers and, thus, interstate conflicts. In the American West, questions of allocation typically dominate interstate water disputes. The Constitution provides two strategies for resolving these conflicts (Getches, 1990).[7] First, as the holder of 'original jurisdiction' in disputes among states, the United States Supreme Court is empowered to resolve interstate complaints. Traditionally, this has been done using the highly flexible doctrine of 'equitable apportionment' in which issues of equity and need are used to craft allocations that can be later revisited by the court should conditions change. The initial use of equitable apportionment was on the Arkansas River between Colorado and Kansas in 1907, although the most celebrated case in 1931 concerned the Delaware River.[8]

The second and much more common approach for resolving interstate conflicts in the West has been the use of interstate compacts (McCormick, 1994). Compacts are legally binding agreements between states, as authorized by the compact clause of the Constitution. States generally prefer compacts over equitable apportionment proceedings since they can retain control over the dispute resolution process, the terms of the ultimate agreement, and the implementation arrangements. Compacts also allow allocations to occur long before needs materialize, which can greatly aid long-term planning and management programs. For these and other reasons, even the courts typically encourage compacts over judicial proceedings.[9]

Interstate compacts can be found throughout western river basins and the plains to the east receiving Rocky Mountain snowmelt. Examples include the Arkansas (Colorado–Kansas, 1949; Kansas–Oklahoma, 1965; and Arkansas–Oklahoma, 1970), Bear (Idaho–Utah–Wyoming, 1955), Belle Fourche (Wyoming–South Dakota, 1943), Big Blue (Nebraska–Kansas, 1971), Canadian (New Mexico–Texas–Oklahoma, 1950), Colorado (Wyoming–Colorado–Utah–New Mexico–Nevada–Arizona–California, 1922), Costilla Creek (Colorado–New Mexico, 1944), Klamath (Oregon–California,

1956), La Plata (Colorado–New Mexico, 1922), Pecos (New Mexico–Texas, 1949), Red (Texas–Oklahoma–Arkansas–Louisiana, 1978), Republican (Colorado–Nebraska–Kansas, 1943), Rio Grande (Colorado–New Mexico–Texas, 1938), Sabine (Texas–Louisiana, 1953), Snake (Wyoming–Idaho, 1949), South Platte (Colorado–Nebraska, 1923), Upper Colorado (Wyoming–Colorado–Utah–New Mexico, 1948), Upper Niobrara (Wyoming–Nebraska, 1962), and Yellowstone Rivers (Wyoming–Montana–Idaho, 1950). Colorado is a party to nine interstate compacts![10]

Typically, the negotiation and approval of interstate compacts has followed a five-step process: (1) Congress authorizes the states to negotiate a compact; (2) state legislatures appoint commissioners; (3) the commissioners meet, usually aided by a federal chairman, to negotiate and sign the agreement; (4) the state legislatures ratify the compact, and (5) Congress ratifies the compact. Omitted from this description is the role of federal water development in stimulating agreements, as the Department of the Interior typically required states to resolve interstate water allocation disputes prior to commencing federally funded river basin developments. The best example of this phenomenon occurred in the Upper Colorado River Basin, where a Bureau of Reclamation study identifying 134 potential projects prompted the basin states within four months to begin compact negotiations (Terrell, 1965).

The key element in interstate water allocation compacts – and for that matter, many international treaties – is the mathematical formula used to apportion flows. Four different allocation strategies are typically seen: (1) systems based on maintaining minimum flow levels at state lines (or other useful gauging stations); (2) approaches based on reservoir storage; (3) formulas allocating fixed or percentage-based rights to consumption or diversion, and (4) a requirement – seen only in the Colorado River basin – for upstream states to deliver downstream a minimum *volume* (rather than a constant *flow* rate) over a lengthy time period. Several formulas have been problematic, largely due to incorrect assumptions about precipitation and run-off levels, a failure to consider surface water/groundwater connections, and due to the growth of water demands in some areas beyond compact apportionments (Kenney, 1996).

Administering compact allocations and resolving conflicts are duties frequently delegated to compact commissions formed by the interstate agreements. Most compacts feature a compact commission, often with a federal (usually non-voting) member. In many cases, however, disputes escalate to the judiciary. Among the most problematic compacts have been those for the La Plata, Pecos, Canadian, Arkansas, Rio Grande, and Colorado Rivers.

Water allocation compacts often provide an element of certainty, stability, and civility in interstate water issues. Ironically, this certainty can be somewhat counterproductive, in that it can eliminate the need and opportunity for continued interaction among the basin states. With the very limited exception of periodic meetings of compact commissioners, so-called 'successful' compacts generally do not require interstate coordination or ongoing cooperation, and provide little reason for one state to be concerned with the water needs of the other. Unlike an equitable apportionment, compacts cannot be modified unilaterally except, perhaps, by congressional action – and no congress has demonstrated an interest in testing that power.

Compacts also do not effectively reconcile hydrologic and political regions. While the signatories to a compact may collectively encompass the entire drainage basin of a particular river, the boundaries of those states do not follow the actual contours of the river basin. Consequently, within states, issues arise about whether to use compact apportionments within the basin itself, or in areas outside the basin. Many of the largest users of the Colorado River, for example, lie outside the topographic bounds of the river basin, but are within the states recognized in the compacts. Similarly, most compacts fail to address water rights associated with tribal lands and other federally reserved lands within the signatory states.

Also of concern in most compacts is the limited attention given to competing water uses and sectors, and in the case of environmental protection, competing water values. With few exceptions, these issues are dealt with in the context of state water law, often with the use of markets.[11] One of the few exceptions is the Northwest Planning Power Council, which is charged with balancing hydropower generation and salmonid management in the United States section of the Columbia River system.[12] This sort of multifaceted mandate is rarely seen in western compacts and compact commissions; however, in the Midwest and East, interstate arrangements addressing pollution control, flood control and planning, and project development are relatively common (Muys, 1971).

Local Issues and the Importance of Scale

Multiscalar studies are on the cutting edge of research in water resources management. Much literature on transboundary waters treats political entities as homogeneous monoliths: 'Canada feels…' or 'The United States wants…' Analysts are only recently highlighting the pitfalls of this approach, often by showing how different subsets of actors relate very different 'meanings' to water (see, for example, Blatter and Ingram, 2001). Rather than being simply another environmental input, water is regularly treated as a security issue, a gift of nature, or a focal point for local society. Disputes, therefore, need to be understood as more than 'simply' over a quantity of a resource, but also over conflicting attitudes, meanings, and contexts. In the American West, local water issues revolve around core values which often date back generations. Irrigators, Native Americans, and environmentalists, for example, can see water as tied to their very ways of life, and increasingly threatened by newer uses for cities and hydropower.

This shift means that water management must be understood in terms of the specific, local context. History matters, as do power flows – the 'meaning' of water to its users is as critical to understanding disputes, and sometimes more so, than its quantity, quality, and timing. For this new world, new tools for analysis are being added to the traditional arsenal, including network analysis, discourse analysis, and historical and ethnographic analysis, each of which can be bolstered and made more robust through the judicious application of appropriate information technologies.

One highlight of these new approaches is that the results of conflict analysis are very different depending on the scale being investigated. To clearly understand the dynamics of water management and conflict potential, then, thorough assessments would investigate the dynamics at multi-scales simultaneously. María Rosa García-Acevedo (2001), for example, puts nominally a 'United States–Mexico' dispute over the Colorado into its specific historic context, and tracks water's changing meanings to the local populations involved, primarily indigenous groups and United States and Mexican farm communities, throughout the 20th century. The local setting strongly influences international dynamics and vice versa.

Similarly, it can be equally useful to follow the dynamics of an issue as it grows in scope and political geography. For example, water resources issues in the Columbia River basin transitioned from intranational to international in 1944 as Canadian and United States planners recognized that cooperative development might well be superior to individual actions, and both countries requested the International Joint Commission (IJC) to study the feasibility of cooperative development in the Columbia Basin. By 1964, the Columbia River Treaty and Protocol were ratified by the governments of Canada and the United States. The treaty is one of the most sophisticated in the world, particularly because it circumvents the zero-sum approach to allocating fixed quantities of water by instead allocating to each country an equal share of benefits derived from the shared basin. Hydropower production, flood control, and other benefits are quantified and shared annually, and there is little dispute across international boundaries.

Many water issues in the Columbia basin and elsewhere, however, have defied a centralized approach. In response to the weaknesses of top–down legislation over locally generated issues such as non-point source pollution and salmon habitat restoration, management authority in the Columbia has been steadily diffusing to local watershed councils. This trend is seen throughout the West and, more sporadically, in several other nations, with mixed success. This trend further reinforces the value of considering multiple scales in defining and addressing transboundary water issues.

2. INTERNATIONAL WATERS: CONFLICT AND COOPERATION

Threat of Conflict

Only 2.5 per cent of the world's water is fresh water, and only a small fraction of that amount is readily available for human use. This renewable but not infinite resource is becoming increasingly scarce. The amount available to the world today is almost the same as it was when man last went to war over it in Mesopotamia some 4500 years ago. At the same time, global demand has steadily increased. In the last 50 years, world population has swelled from 2.5 billion to 6 billion, while the renewable supply per person has fallen 58 per cent (Wolf et al., 2003). Moreover, unlike oil and most other strategic resources, fresh water has no substitute in most of its uses. It is essential for growing food, manufacturing goods, and safeguarding human and environmental health. And while history suggests that cooperation over water is the norm, it is not the rule.

United Nations Secretary-General Kofi Annan is on record as warning that 'fierce competition for fresh water may well become a source of conflict and wars in the future' (Postel and Wolf, 2001), and a recent report of the United States National Intelligence Council concludes that the likelihood of international conflict will increase during the next 15 years 'as countries press against the limits of available water' (NIC, 2000, p. 27). These conflicts are likely to go beyond traditional issues of surface water flows to include groundwater depletions and, increasingly, the long-neglected issue of water quality.

The greatest threat of the global water crisis comes from the fact that people and ecosystems around the globe lack access to sufficient quantities of water at sufficient quality for their well-being. The concern over water is spreading as populations increase. By 2015, nearly 3 billion people – 40 per cent of the projected world population – are expected to live in countries that find it difficult or impossible to mobilize enough water to satisfy the food, industrial, and domestic needs of their citizens. This scarcity will translate into heightened competition for water – both within and between political regions.

The largest and most combustible imbalance between population and available water supplies will be in Asia, where crop production depends heavily on irrigation. Asia today has roughly 60 per cent of the world's people but only 36 per cent of the world's renewable fresh water. China, India, Iran, and Pakistan are among the countries where a significant share of the irrigated land is now jeopardized by groundwater depletion, scarce river water, a fertility-sapping buildup of salts in the soil, or some combination of these factors. Groundwater depletion alone places 10 to 20 per cent of grain production in both China and India at risk. Water tables are falling steadily in the North China Plain, which yields more than half of China's wheat and one-third of its corn, as well as in northwest India's Punjab, another major breadbasket.

As farmers lose access to irrigation water and see their livelihoods deteriorate, they may not only resort to violent protest but migrate across borders to restive and already overcrowded cities. Such has been the case in Pakistan, where falling agricultural output has prompted a massive rural migration to large urban centers, leading to renewed outbreaks of ethnic violence.

Mechanism of Conflict[13]

The scarcity of water leads to intense political pressures, often referred to as 'water stress'. Water shortages have contributed to tensions between competing uses around the globe, pitting town against country, environment against industry, and upstream users against downstream interests. One-fourth of water-related interactions during the last half-century were hostile. Although the vast majority of these hostilities were confined to verbal antagonism, rival countries took to arms on 37 recorded occasions.

Most transboundary water conflicts share a common trajectory. A three-year study of conflict and cooperation within international river basins by researchers at Oregon State University found that the likelihood of conflict increases significantly whenever two factors come into play (Wolf et al., 2003). First, some large or rapid change occurs in the basin's physical setting (typically the construction of a dam or other development project) or in its political setting, especially the breakup of a nation that results in new international rivers. Second, existing institutions are unable to absorb and effectively manage that change, due to the lack of a treaty spelling out each nation's rights and responsibilities with regard to the shared river or implicit agreements or cooperative arrangements. Even technical working groups can go some way to managing contentious issues, as they have in the Middle East.

Over the next ten years, 17 river basins appear ripe for tension or conflict; in another four, serious unresolved water disputes already exist or are being negotiated. These basins encompass 51 nations on five continents in just about every climatic zone. Consider, for example, the Salween River, which rises in southern China, then flows into Myanmar (Burma) and Thailand. Each of these nations plans to construct dams and development projects along the Salween, and no two sets of plans are compatible. To compound matters, China was one of just three countries that voted against a 1997 UN convention that established basic guidelines and principles for the use of international rivers.

Other river basins are at risk of disputes due to rapid political changes. The breakup of the Soviet Union resulted in several new international river basins almost overnight, and, not surprisingly, institutional capacity for managing water disputes in them is weak. The watershed of Central Asia's Aral Sea, for instance, spanned five Soviet republics that are now independent countries. Tensions among the young nations quickly arose both over how to share the Amu Dar'ya and Syr Dar'ya, the two rivers that feed the Aral Sea, as well as how to ameliorate the human and environmental tragedy caused by the sea's dramatic shrinking – a result of 40 years of river diversions masterminded by Moscow to grow cotton in the Central Asian deserts. With assistance from international agencies, these young governments have taken tentative steps toward trying to resolve their water dilemmas.

Cooperation is Key

There is certainly room for optimism, though. Historically, cooperation typi-cally prevails over conflict. The most vehement enemies around the world, whether Israelis and Arabs, Indians and Pakestanis, or Armenians and Azeris, either have negotiated water-sharing agreements, or are striving to realize them. Violence over water seems neither strategically rational, hydrographically effective, nor economically viable. Shared interests consistently outweigh water's conflict-inducing characteristics.

Looming crises in food production may also stimulate new international alliances, as countries with increasingly deficient water supplies for food production seek new and expanded partnerships with grain exporters. For example, Asia, Africa, and the Middle East already account for 26 per cent of global grain imports, and as these water-stressed regions add an additional billion people over the next 15 years, may be compelled to form stronger alliances with potential exporters in China, India, and Pakistan.

Of course, for those nations without sufficient foreign exchange to turn to imports, notably those in sub-Saharan Africa, higher world grain prices will likely mean greater hunger, political insecurity and conflict, and more calls for humanitarian aid. The challenge for the international community is to get ahead of the 'crisis curve', to help develop institutional capacity and a culture of cooperation in advance of costly, time-consuming crises, which threaten lives, regional stability, and ecosystem health.

Selected Case Studies[14]

Hundreds of examples of transboundary conflict and cooperation exist through-out the world. While several examples have already been referenced, some particularly salient cases are discussed below in greater detail, with the aim of further exploring themes already mentioned while identifying new find-ings and lessons that may have applicability to the American West.

Mekong River of southeast Asia

The Mekong Basin, which includes China, Myanmar, Thailand, Cambodia, Vietnam, and Laos is an example of an international basin that has been successfully managed through cooperation of a shifting set of riparians (Jacobs, 2002). Since 1957, coordinated water resources planning in this basin has been the goal of the international institution now named the Mekong River Commission (MRC). The MRC and its predecessors have used legal and organizational tactics to promote cooperation between riparian nations. Laws require unanimous consent for mainstem projects, maintaining focus on basinwide management. Although diversions from the Mekong are legal, the lack of apportionment between riparians has also contributed to a coopera-tive, basinwide planning approach. These laws, in turn, have upheld the resiliency of the institution.

The Mekong River Commission has focused on scientific investigations and data collection of basin hydrology, ecosystems, and human and legal aspects of water management. This science-based approach of the MRC has encouraged collaboration between otherwise rivaling parties. Additionally, the MRC has recently made a transition from a project-oriented approach to a program-oriented approach. Furthermore, the lack of large development on the Mekong River has allowed the MRC to keep its options for development broad. Large upstream dams could threaten this, however.

The Mekong River Basin serves as an example where an international water resource has helped to unite, rather than divide, riparians. Riparian nations foresaw more potential benefits through multilateral cooperation than through unilateral approaches to development on the Mekong. Flexibility in river management and institutional capacity have been keys to cooperation, and the prospect of donor aid has provided a significant incentive for coop-eration. The exemplary actions of the Mekong River Commission and its experience in conflict resolution and collaborative science programs could help inform interstate and interbasin dialogue in the American West. The shift of priority of the MRC from projects to programs resembles the United States Bureau of Reclamation's shift from water development to water manage-ment, and both institutions could learn from the other's shift.

Okavango River in southern Africa

An examination of the Okavango River in southern Africa yields many valuable lessons for water management in the western United States (Hitchcock, 2002). A major international river, the Okavango provides important natural resources and ecological functions as well as ideological significance to the populations who live in the basin and use the resources within it. Knowledge of rights concerning riverine land and resources is passed on from one generation to the next. Historically, the socioeconomic values of riparian resources have been determined by the local population, and the people in turn have had the right to handle the resources however they so choose. However, changes have been occurring recently in the basin. Cattle owners and safari companies have been increasing in numbers, and a trend toward privatization of resources is evident. These changes are linked to rising populations, increasing development, environmental degradation, and conflicts over resource use. These issues are exacerbated by the increasingly undefined status of land and the lack of secure authority structures. This changing situation coupled with the dire shortages of water that are projected for the next few decades make the future tenuous.

Angola, Namibia, and Botswana, the major riparians of the Okavango, all have different national administrative structures at the national and district levels that handle water resources management, but they are held together with the regional institutional structure of the Permanent Okavango River Basin Commission, formed in 1994 by these riparian nations. The Trinational Permanent Water Commission, also formed by these three nations, provides advice on environmentally and socially sustainable development of the Okavango River waters. The establishment of this interbasin form of governance has helped to alleviate pressures that lead to periodic outbreaks of conflict regarding the Okavango.

Many lessons have come out of the experience of these institutions that can be applied to resource planning the western United States. The involvement of local populations in decision making regarding natural resources can improve the planning process on all scales. The existence of international (or in the case of the western United States, also interstate or interbasin) institutions can aid greatly in resolving conflicts over resources, harmonizing water resource policies, and monitoring compliance. A protocol established by such an institution can be valuable if members states are willing to respect the protocol and operate by its rules. Finally, the implementation of social and environmental justice principles in all policies and practices is critical to the resolution and prevention of conflicts.

The West Bank

A complex set of interactions over many scales from local to international influence water resources in the West Bank (Trottier, 2002). In the highly politicized Israeli–Palestinian conflict in the West Bank, Israel, the stronger negotiating party, hands down military orders regarding water allocations from aquifers. Palestinians in the West Bank are allocated less than one fourth the amount that Israelis receive, and pumping more water from Palestinian wells has come to be viewed as a nationalist act, threatening the resource. However, beyond this allocation, Israel interferes little with the management of water resources.

Although irrigation is not well developed in the West Bank, 65 per cent of water goes toward irrigation practices. Irrigation communities often have communal property regimes for water management. Water for irrigation mostly comes from springs and wells, each of which has different regimes of management. Water from springs is often allocated based on nested local institutions that can include, for example, families, groups, and subgroups within a village, that are in turn connected to a reticulation network operated by the municipality. Wells, which have appeared in the last several decades in the West Bank, are managed through private bargaining of farmers and well owners. Water management for agriculture is handled informally and is independent of the Palestinian Authority. These institutions of water control are very resilient and are involved in many facets of water control.

The management of water for domestic use stands in contrast to that of agricultural use. The arrival of the Palestinian Authority brought about many changes in domestic allocation from its traditional local management. The Palestinian Water Authority is involved in providing water for domestic use, although it faces challenges from local forces as well as the Ministry of Local Governments, which currently controls many reticulation networks. In addition, many communities are not yet connected to reticulation systems, but receive supplies from water tankers which have control over their own pricing structures.

The juxtaposition of the traditional view of water resources in terms of social control and security with the newer view of water being a public good plays a critical role in the difficulties of managing water in the West Bank and presents a challenge to Palestinians. In order for any approach towards sustainable management of water resources to be effective, both views have to be taken into account. All dimensions of interactions must be considered. This provides a valuable insight for water management in the American West. Although the power configurations regarding water in the western United States and the West Bank are different, multiscalar analyses of the impact of various policies can prove valuable to both.

3.   FINDINGS AND CONCLUSIONS

This global review of transboundary water conflicts yields a variety of insights. Three findings are particularly prominent:

1. Water crossing international (and interstate) boundaries often causes tensions between nations (and states) that share the basin. Early coordination between riparian governments, however, can often help ameliorate conflict.
2. Once international institutions are in place for transboundary water resources, they prove tremendously resilient, even as conflict is waged over other issues.
3. As transboundary water disputes evolve, a gradual decrease in water quantity or quality is a more likely outcome than violent conflict. Over time this can affect the internal stability of a nation or region. The resulting instability may have effects in the international arena.

Transferring Lessons: Key Considerations

Converting findings (observations) to transferrable lessons (practical advice) is a significant challenge. It should be clear from the cases presented in this study that both similarities and distinct differences are inherent between regions of the world, and between national (including interstate) and international water conflicts. This provides both opportunities and constraints to the transference of international findings to the transboundary conflicts of the American West. Four key issues to consider when evaluating the transferability of case study findings are listed below.

1.   Institutions and authority

National cases often are played out in relatively sophisticated institutional settings – such as found in the American West – while international conflicts can be hampered by the lack even of an institutional capacity for conflict resolution. Presumably this would make national (including interstate) conflicts more amenable to resolution, however, it can be argued that the presence of strong law-dominated approaches to resolving conflicts can impede creative problem solving, effectively presenting the same challenges as the international setting.

2.   Law and enforcement

A similar dichotomy exists regarding issues of legal enforcement of agreements. Over the years the United States and other countries have established intricate and elaborate legal structures to provide both guidance in cases of domestic water disputes, and a setting for clarifying conflicting interpretations of that guidance. International disputes, in contrast, rely on poorly defined water law, a court system in which the disputants themselves have to decide on jurisdiction and frames of reference before a case can be heard, and little in the way of enforcement mechanisms. (One result is that international water conflicts are rarely heard in the International Court of Justice. Likewise, of the international cases presented in this volume, only the Mekong Committee has used the legal definition of 'reasonable and equitable' use in its agreement.) Nonetheless, it has been argued that the differences between national and international disputes are more apparent than real, and that given the myriad of legal venues open to disputants (in national settings), and ambiguities of court jurisdiction, creative lawyers can effectively hamstring legal challenges for years, essentially creating a de facto lack of legal authority.

3. Presumption of equal power

'All are equal in the eyes of the law' is a common phrase describing national legal frameworks. Yet, no such presumption exists in international conflicts, where power inequities define regional relations. Most of the international watersheds presented here include a hegemonic entity which brings its power to bear in regional negotiations, and which often sees agreements tilt in its favor as a consequence. Yet, to attribute such inequities just to international settings is tenuous, as financial or political inequities in a national setting can also ensure that conflict resolution processes lead to distorted outcomes.

4. The Best Alternative to a Negotiated Agreement (BATNA)

A difference commonly pointed out between national and international disputes is that, in national water conflicts, war is not usually a realistic alternative to a negotiated agreement – what the dispute resolution profession terms a 'BATNA' (Best Alternative to a Negotiated Agreement). While it may be true that intranational 'water wars' are not likely, the same is increasingly accepted as being true of the international setting. While shots have been fired, both nationally and internationally, and troops have been mobilized between countries, no all-out war has ever been caused by water resources alone. As one analyst familiar with both strategic issues and water resources has noted, 'Why go to war over water? For the price of one week's fighting, you could build five desalination plants. No loss of life, no international pressure, and a reliable supply you don't have to defend in hostile territory' (Wolf, 1998, p. 261).

Three Lessons for the American West

Effectively addressing the issues raised by transboundary resources will be a chronic problem in the American West. Nonetheless, three lessons gained from international experience appear to offer real promise for improved outcomes.

1. Seek flexible solutions based on needs, not rights

What one notices in the global record of water negotiations is that several of those surveyed begin where many western United States issues are now, with parties basing their initial positions in terms of rights – that is, the sense that a riparian is entitled to a certain allocation based on hydrography or chronology of use. Irrigators in the Klamath basin, for example, invoke rights under the Reclamation Act while environmentalists refer to the Endangered Species Act. Upstream riparians often invoke some variation of the Harmon Doctrine, claiming that water rights originate where the water falls. Downstream riparians, meanwhile, often claim absolute river integrity, claiming rights to an undisturbed system or, if on an exotic stream, historic rights based on their history of use.

In almost all of the disputes globally which have been resolved, however, particularly on arid or exotic streams, the paradigms used for negotiations have not been 'rights-based' at all – neither on relative hydrography nor specifically on chronology of use – but rather 'needs-based'. 'Needs' are defined by irrigable land, population, or the requirements of a specific project.

Similarly, successful frameworks in the international experience are flexible; flexibility in agreements is almost more critical than the initial agreements themselves. Needs, interests, flow regimes, and values all change over time. But the record shows that human creativity, when applied with a modicum of goodwill, has overcome these obstacles in settings plenty more complex and hostile than the western United States, which should come as quite encouraging news for our area of interest.

2. Expand the pie

Most of the treaties and interstate agreements reviewed deal exclusively with water (and typically just water allocation), separate from any other political or resource issues between countries – water qua water. By separating the two realms of 'high' and 'low' politics, or by ignoring other resources which might be included in an agreement, some have argued, the process is either likely to fail, as in the case of the 1955 Johnston accords on the Jordan, or more often to achieve a suboptimum development arrangement, as is currently the case for the Indus agreement, signed in 1960. Increasingly, however, linkages are being made between water and politics, and between water and other resources. These multiresource linkages may offer more opportunities for creative solutions to be generated, allowing for greater economic efficiency through a 'basket' of benefits. Some goods that have been included in water negotiations include financial resources, energy resources, political linkages, transportation infrastructure, and data.

Policy makers may want to consider a change in focus from the current water-specific concentration to an aggregate view of river basin resources. Coriparians may find it beneficial, for example, to pool water allocation issues with other river basin projects. A bundling of river basin resources may not only provide additional bargaining options, but may also, by reducing duplicative efforts, result in a more efficient and mutually beneficial allocation of resources, both natural and monetary. For example, a downstream riparian might offer financial assistance for a hydroelectric project in exchange for some percentage of power, or an upstream riparian might support the construction of locks and dams in a downstream riparian state in exchange for navigation rights (Krutilla, 1969).

Limited precedents for multi-purpose linkages currently exist in certain international water treaties. The 1964 United States–Canada Columbia River treaty, for example, allocates water according to an equal distribution of benefits, defined by hydropower generation and flood control. (Incidentally, this results in the odd arrangement that power may be exported out of basin for gain, but the water itself may not.) Similarly, as part of the 1975 Mekong River Agreement, Thailand provided financial support for a hydropower project to Laos in exchange for a percentage of the electricity generated. India and Nepal also bundled projects such as irrigation, hydropower, navigation, fishing, and aforestation into two treaties concluded in the 1950s and 1960s. Other resources that have been or could potentially be included in water negotiations include data, technology, and political capital (Wolf, 1998).

3. Devise institutions for ongoing coordination

In managing transboundary water resources, strong institutions make a difference. The Indus Waters Treaty, for example, survived two wars between the signatories and allowed each to pursue its agricultural and economic plans without risking the ire of the other. Long-term programs of joint fact finding, technical cooperation, and other initiatives that establish a climate of cooperation among countries can pave the way for resolving disputes when they do arise. The Global Alliance for Water Security aimed at coordinating assistance in priority regions may help countries get ahead of the crisis curve.

But what should a basinwide institution look like? Despite the tendency of water managers to think in terms of total integration of watersheds, this often is not the most likely or practical outcome. Even friendly nations internationally often have difficulty relinquishing sovereignty to a supralegal authority, and the obstacles only increase along with the level of suspicion and rancor. Consequently, one should strive for coordination over true integration. Once the appropriate benefits are negotiated, it then becomes an issue of 'simply' agreeing on a set quantity, quality, and timing of water resources that will cross each border. Coordination, when done correctly, can offer the same benefits as integration, and be far superior to unilateral development, but does not threaten the one issue all states hold dear: their sovereignty.

NOTES

1. Treaty Relating to Boundary Waters Between the United States and Canada, 11 January 1909, 36 Stat. 2448.
2. Treaty between Canada and the United States of America Relating to Cooperative Development of the Water Resources of the Columbia River Basin, ratified by the US Congress on 16 March 1961, and by Canada three years later.
3. Treaty between the United States and Mexico Respecting Utilization of Waters of the Colorado and Tijuana Rivers and of the Rio Grande, 3 February 1944, 59 Stat. 1219.
4. Treaty of Peace, Friendship, Limits, and Settlement with the Republic of Mexico, 2 February 1948, United States and Mexico, 9 Stat. 922.
5. Convention between the United States of America and the United States of Mexico, 1 March 1889, art.1, 26 Stat. 1512.
6. 'The fundamental principle of international law is the absolute sovereignty of every nation, as against all others, within its own Territory' (cited in LeMarquand, 1993, p. 63). Harmon was making the hydrologically preposterous argument that upstream water diversions within the territorial US would not legally affect downstream navigation on international stretches of the Rio Grande since the diversions were to be carried out by individuals, not States (McCaffrey, 1996).
7. Arguably, a third strategy also exists: congressional apportionment. This approach is not included here because it has only been observed in one, highly unusual situation, and is generally not expected to emerge again as a means for interstate apportionment. The case in question involved allocation of the Lower Colorado River among Arizona, California, and Nevada, something that Congress effectively did (according to a later court decision) in the Boulder Canyon Project Act of 1928 (Getches, 1990).
8. *Kansas v. Colorado*, 206 U.S. 46 (1907); *New Jersey v. New York*, 283 U.S. 336 (1931).
9. For example, *see Colorado v. Kansas*, 320 U.S. 383, at 392 (1943).
10. Interstate water allocation compacts are becoming fashionable in the East, as found in the Delaware, Susquehanna, Apalachicola-Chattahoochee-Flint (ACF), and Alabama-Coosa-Tallapoosa (ACT) River Basins. The Delaware and Susquehanna compacts are unique in that they involve the federal government as a signatory and partner (so-called federal-interstate compacts) (GAO, 1981). The agreements in the ACT/ACF basins are unique in that they do not include allocation formulas, but rather establish commissions empowered to later devise allocation compacts.
11. Interstate water markets have not materialized, and may not be legally viable under many compacts.
12. The Northwest Power Planning Council is, admittedly, an odd arrangement led by appointees from the four basin states, formed by a combination of interstate compact and federal legislation, and charged primarily with regulating federal activities – New Federalism in the extreme (Volkman and Lee, 1988).
13. This section is drawn primarily from Postel and Wolf (2001).
14. Thanks to Kristin Anderson for helping to compile these summaries.

LITERATURE CITED

Blatter, Joachim, and Helen Ingram (eds) (2001), *Reflections on Water: New Approaches to Transboundary Conflicts and Cooperation*, Cambridge, MA: MIT Press.
General Accounting Office (GAO) (1981), 'Federal-interstate compact commissons: useful mechanisms for planning and managing river basin operations', report to the Congress by the Comptroller General of the United States, 20 February.
Garcia-Acevedo, Maria Rosa (2001), 'The confluence of water, patterns of settlement, and constructions of the border in the Imperial and the Mexicali Valleys (1900–1999)', in Joachim Blatter and Helen Ingram (eds), *Reflections on Water: New Approaches to Transboundary Conflicts and Cooperation*, Cambridge, MA: MIT Press, pp. 57–88.
Getches, David H. (1990), *Water Law in a Nutshell*, St Paul MN: West Publishing Co.
Hitchcock, Robert K. (2002), 'The Okavango River Basin in southern Africa: a case study of transboundary resource management issues', paper prepared for the Natural Resources Law Center, University of Colorado School of Law, Boulder, CO.
Holburt, Myron S. (1975), 'International problems on the Colorado River', *Natural Resources Journal*, **15**, 11–26.
Jacobs, Jeffrey (2002), 'Transboundary river management in the Mekong River Basin: key issues and lessons for western U.S. water management', paper prepared for the Natural Resources Law Center, University of Colorado School of Law, Boulder, CO.
Kenney, Douglas S. (1996), 'Review of coordination mechanisms with water allocation responsibilities', in *Phase 2 of Coordination Mechanism Research for the ACT-ACF Comprehensive Study: Final Report*, Carbondale IL: Planning and Management Consultants, Ltd, pp. 151–68.
Krutilla, John V. (1969), *The Columbia River Treaty – The Economics of an International River Basin Development*, Baltimore, MD: Johns Hopkins University Press.
LeMarquand, David (1993), 'The International Joint Commission and changing Canada–United States boundary relations', *Natural Resources Journal*, **33**(1), 59–92.
McCaffrey, Stephen C. (1996), 'The Harmon Doctrine one hundred years later: buried, not praised', *Natural Resources Journal*, **36**(3), 549–90.
McCormick, Zachary L. (1994), 'Interstate water allocation compacts in the western United States – some suggestions', *Water Resources Bulletin*, **30**(3), 385–95.
Mumme, Stephen (1988), *Apportioning Groundwater Beneath the United States–Mexico Border: Obstacles and Alternatives*, San Diego: University of California Press.
Mumme, Stephen P. (1999), 'Managing acute water scarcity on the U.S.–Mexico border: institutional issues raised by the 1990s drought', *Natural Resources Journal*, **39**, 149.
Muys, Jerome C. (1971), *Interstate Water Compacts*, National Water Commission publication no 202998, Washington, DC: US Government Printing Office.
National Intelligence Council (NIC) (2000), *Global Trends 2015*, Washington, DC: Director of Central Intelligence.
Postel, Sandra L. and Aaron T. Wolf (2001), 'Dehydrating conflict', *Foreign Policy*, September/October, 60–7.
Terrell, John Upton (1965), *War for the Colorado River: Volume Two, Above Lee's Ferry*, Glendale, CA: The Arthur H. Clark Company.

Trottier, Julie (2002), 'Case study – the West Bank', paper prepared for the Natural Resources Law Center, University of Colorado School of Law, Boulder, CO.

Volkman, John M. and Kai N. Lee (1988), 'Within the hundredth meridian: western states and their river basins in a time of transition', *University of Colorado Law Review*, **59**, 551.

Wolf, Aaron (1998), 'Conflict and cooperation along international waterways', *Water Policy*, **1**(2), 251–65.

Wolf, Aaron T., Shira B. Yoffe and Mark Giordano (2003), 'International waters: identifying basins at risk', *Water Policy*, **5**(1), 29–60.

6. Sustainability and the future of western water law

Lakshman Guruswamy and A. Dan Tarlock

INTRODUCTION

The chapters in this book clearly demonstrate that the American West is not the only region on the planet challenged to manage better its scarce water. The problems of policy and law confronting the American West are common to other parts of the world. The question is: what are the appropriate institutional models and legal principles to address them? Unfortunately, the answer is not as easy as it once seemed. At one time, the United States was the undisputed leader in water policy. It initially adapted the experience of the British Empire and Spain to the arid and semi-arid West and used this experience to build the institutions necessary to sustain the settlement of the region (*see* Chapter 1). But, the chapters in this volume also demonstrate that the United States is lagging behind other parts of the world in adapting to changed water use demands. The old model of optimum river basin development no longer encompasses the full range of economic, environmental and social dimensions of modern water use. We want water to serve traditional consumptive and non-consumptive demands, to support aquatic ecosystems and to help sustain defined communities such as indigenous peoples. Most countries of the world are attempting to apply an expanded version of sustainable development to construct a new water use policy.

We argue that there is a need to construct a new western water use policy and that it should reflect the international consensus about the universally adopted parent concept of Sustainable Development (SD) and its offspring, Integrated Water Resources Management (IWRM). IWRM has attempted to grapple with difficulties similar to those that tax the American West caused by multiple users and multiplying uses of water, along with their varying and unmet demands. Sustainable water use in the American West cannot satisfactorily be achieved through the presently fragmented and fractured overlay of policy and law, but could more efficiently and fairly be undertaken within the internationally offered framework of Integrated Water Resources Management (IWRM). In the United States, and the American West, the most promising opportunity for implementing IWRM is through the evolving concept of Integrated Watershed Management (IWM). Admittedly, IWM will not be easy to implement because it confronts a number of political, social, and economic problems arising out of the change from the fragmented and piecemeal administration of law and policy, to a comprehensive and integrated approach. These complex questions do not admit of answers within the modest page limits and objectives of this chapter. Instead, we focus on the more limited but important question of the possible legal constraints on the adoption and practice of IWM.

Integrated Watershed Management faces at least two legal challenges. First: to what extent does the law confer jurisdiction or legal authority upon any agency to create, coordinate and implement a comprehensive and integrated approach? Second: to what extent might an integrated and comprehensive approach require the modification of pre-existing entitlements to water? This chapter offers an introduction to these two problems.

Before describing the structure of this chapter it may be helpful to offer a preliminary basis for distinguishing three interrelated concepts (and their acronyms). First, Integrated Water Resources Management (IWRM) is an international construct most clearly articulated in Agenda 21 of the United Nations Convention on Environment and Development held in Rio de Janeiro in 1992, and reaffirmed at the World Summit on Sustainable Development held in Johannesburg in 2002. Second, Integrated Watershed Management (IWM) is a United States concept that closely resembles IWRM. The third conceptual framework of Unified River Basin Management (URBM) is discussed because IWM was derived and evolved from it. URBM is of historical rather than contemporary importance.

Part 1 outlines the issues confronting western water law, thus setting the factual and policy scene calling for a different response to what has traditionally been offered. Parts 2 and 3 explain the international response being fashioned through the concepts of Sustainable Development and Integrated Water Resources Management. Part 4 traces the development of the American concept of Integrated Watershed Management, delineating the extent to which it expresses IWRM, and advances the case for its implementation. Part 5 addresses some of the legal difficulties of reapportioning water rights or entitlements that will become necessary if IWM is adopted.

1.   CHALLENGES CONFRONTING WESTERN WATER LAW

It is difficult to address western water problems in an integrated way because we have long assumed that the problem was one of supply augmentation and that new demands, such as environmental quality, could be addressed through marginal adjustments without disturbing fundamental supply augmentation and conservation programs (see Chapter 3). There has always been competition for the West's unevenly distributed and often scarce water resources. Until recently, the competition has been largely masked by federal and state policies to subsidize large carry-over storage reservoirs and by the widespread toleration of groundwater mining. The three major user groups – irrigated agriculture, municipal and urban users and hydroelectric power generators – were able to share the region's available water resources with a minimum of friction, although many states, such as California and Colorado, experience fierce political and legal battles among these user groups. But, until the 1980s, the federal government and the states were able to practice successfully the politics of distribution.

Today, the competition is intensifying as most of the region continues to attract continued in-migration, and long-excluded uses, such as those of Native Americans and the environment, demand a share of the resources. Not only are there increasing demands on already stressed resources, but the competition is taking place in a radically changed institutional and perhaps hydrologic environment (see Chapter 2). We believe that these pressures and the new institutional and hydrological environment will require the adoption of a new ethic of sustainable water use and development. Sustainability principles must inform both the design of new projects and the re-engineering and operation of existing facilities.

The traditional vision of a river system as a commodity to be used to the maximum extent possible is still the dominant vision in many parts of the world such as China and India by choice or necessity, but in countries such as Australia and South Africa it is slowly being replaced by the paradigm of ecologically sustainable development (see Farrier, 2002; Stein, 2002; Xi et al., 2002). The basic reason for the paradigm shift is that multiple purpose development imposed substantial environmental costs and increased social inequity in many parts of the world. In recent years there have been sustained efforts to account for these ignored costs. The process of 'environmental accounting' has led to a more radical ecological ideal of managing river systems to maximize ecological services (Teclaff, 1967, 1991). The object is to maintain the river's historic natural services and functions. The newer ecological integrity vision is less clearly articulated than multiple use because it rests on a more complex view of the human role in the functioning of natural systems (Botkin, 1990). It starts from the premise that we must try to integrate human uses of a river system with the maintenance of its natural environmental sustainability both in the design of new projects and the re-engineering and operation of existing facilities (for example, Meyer, 1993, 1994; Noss and Cooperrider, 1994; and Tarlock, 1994).

There is a high degree of recognition among students of water policy and large segments of the water use community that we have reached the end of the big dam era. Future water policy will be guided by three principles: (1) the more efficient use of existing supplies; (2) the use of more sophisticated and less environmentally intrusive technologies to develop new supplies, and (3) the restoration of degraded aquatic ecosystems to maintain and recapture valuable ecosystem services (Bates et al., 1993).[1]

2.   SUSTAINABLE DEVELOPMENT

The definition of Sustainable Development (SD) employed in this chapter calls for the wise use of natural resources such that human needs are fulfilled, but are counterbalanced by the need for resources to continue to be available for use by future generations. SD is a political construct that has been accepted as a foundational norm of environmental law and policy by the international community. In addition, the international community has accepted SD as the groundnorm[2] (basic norm) of international environmental law ever since it was proclaimed to be such at the Earth Summit of 1992,[3] and reaffirmed at the Johannesburg World Summit on Sustainable Development (WSSD) in 2002. Despite its exalted status, the concept of SD remains of a chimerical character that needs to be honed, refined, and more clearly defined. While the concept of Sustainable Development continues to evolve, a recent restatement of SD conceptualized by a group including a significant number of Nobel Laureates is worthy of particular attention. This distinguished group defines SD as the wise use of resources through social, economic, technological, and ecological policies governing natural and human engineered capital.[4] According to this restatement, such policies should promote innovations that assure a higher degree of life support for the fulfillment of human needs while ensuring intergenerational equity.

Such a definition of SD recognizes human interaction with the natural world, and this interaction is consistent with the non-equilibrium paradigm in ecology. The non-equilibrium paradigm may be contrasted with the equilibrium model that calls for preservation of the natural world through the exclusion, or limited intervention, of humans. Thus, the non-equilibrium model integrates humans into the natural world and allows for appropriate human intervention into natural systems in an effort to maximize life support systems.

Traditionally, life support systems have been managed in a fragmented manner. These systems can be better managed if they are viewed as an integrated whole. Scientists have offered illuminating examples of multiple interacting changes affecting water, atmosphere, and biodiversity, among life support systems (Matson, 2001). These examples provide evidence in support of an integrated management approach that embraces not only natural life support systems (physical, chemical, and biological) but also human systems (legal and institutional).

Ecologists are steadfast in suggesting the need for a comprehensive – as distinct from a fragmented – approach to environmental problems. Moreover, by adopting an ecological perspective, scientists offer a better view of the enormous biocomplexity confronting an integrated approach to environmental decision making.

Physical, chemical, and biological scientists, in a remarkable display of interdisciplinary cooperation, confront the interrelated character, or biocomplexity, of global problems by successfully establishing huge billion-dollar initiatives on global change, such as the International Geosphere Biosphere Program and the World Climate Research Program. These efforts present a comprehensive and integrated model of the physical, chemical, and biological processes that regulate the world (Guruswamy and NcNeely, 1998). These studies consider the biosphere, the geosphere, and all the interactions within and between them. For example, studies include 'biogeochemistry' where models track the cycles of chemical elements that flow through ecosystems, 'biogeophysics' where models determine the way energy flows through ecosystems, and additional models that predict the effects of land use on ecosystems. The collection of data for undertaking such an effort has already begun.

Because an integrated approach is a way to achieve both environmental and human goals, its methods of management should be crafted and molded in light of these substantive goals. The exercise of formulating specific substantive goals is an ongoing and dynamic process, informed by the knowledge base emerging out of grassroots, national, regional, and global efforts. In the case of Integrated Water Resources Management or Integrated Watershed Management, these goals must be built on the foundations of sustainable development.

This chapter considers the extent to which ecologicalism – a world view based on the principles of ecology – is anchored in the objectives of Sustainable Development, and leads to the adoption of IWRM and IWM. Ecologicalism establishes a new paradigm about the nature of ecological systems that may call for a review of existing legal objectives. Ecologicalism is based on local, regional, and global interdependency of life support systems, and posits the need for developing integrated management strategies for managing life support systems such as water in a manner that maintains sustainability.

We are currently witnessing a remarkable confluence of politics and science. The politics of SD and the science of the non-equilibrium paradigm are creating conceptions of resource use that were once shunned by equilibrium ecologists, law makers, and policy makers. The convergence of SD and the non-equilibrium paradigm heightens the need for a re-evaluation and redefinition of the objectives, substantive goals, and rationales underlying water management in the United States, as well as in the international arena. For example, The ill-fated Report of the Western Water Policy Review Advisory Commission (WWPRAC, 1998) adopted this definition of sustainable water use: 'Sustainable water resources management builds on the long tradition of state and federal management to conserve water and apply it to a wide range of beneficial uses, but the achievement of sustainability also presents new challenges for which past management practices and institutions often provide limited guidance.' It then identified the core element of a sustainable water policy, concluding that the West needs to 'define hydrologic baselines for individual basins and watersheds that reflect the full range of valued water uses, including ecosystem uses' (WWPRAC, 1998). This is not a simple river preservation concept. Although some aquatic scientists want to subordinate human use to the 'normative' river, the newer river-as-ecosystem concept starts from the premise that we must try to integrate human uses of a river system with the maintenance of its natural environmental services. In short, a sustainable water resource regime is one that strikes a new balance between in and out of stream uses and reduces the economically irrational subsidies that have characterized western water policy.

The concept of Integrated Water Resources Management seeks to apply SD to water resources and it is to that concept that we now turn.

3.   INTEGRATED WATER RESOURCES MANAGEMENT

The need for IWRM was declared by Agenda 21 – the environmental action plan for the 21st century agreed to at the 1992 United Nations Conference on Environment and Development. The case for IWRM was reaffirmed at the World Summit on Sustainable Development (WSSD) held in Johannesburg in 2002.

The WSSD's rearticulation of SD as consisting of three mutually reinforcing pillars – economic development, social development and environmental protection – has implications for IWRM (Johannesburg Declaration on Sustainable Development, Article 5). While WSSD affirmed the importance of economic development established in Rio, it added the third pillar of social development that hitherto was seen as a component of economic development and not as a separate entity. In developing IWRM, WSSD emphasizes the extent to which human needs are fulfilled by environmental protection only to the extent that it is an integral part of economic and social development. While IWRM has always been based upon the foundational premises of SD, the WSSD re-emphasizes the extent to which IWRM must now integrate economic and social development with environmental protection. Consistent with this clearly anthropocentric framework, WSSD warns that adverse human impacts on the integrity of ecosystems might result in impairing vital resources and life support essential for human well-being (WSSD Plan of Implementation, Article 23).

IWRM, as articulated in Agenda 21, calls for the holistic management of freshwater as a finite and vulnerable resource, and the integration of sectoral water plans and programs within the framework of economic and social policy. Water is an economic and social good whose utilization depends on its quality and quantity. IWRM recognizes the importance of water resources in socioeconomic development, and addresses the challenges posed by the competing, multi-interest utilization of water resources for water supply and sanitation, agriculture, industry, urban development, hydropower generation, inland fisheries, transportation, recreation and low and flat land management. The need to utilize water, which forms an integral part of the life support systems of the world, however, does not weaken the extent to which IWRM recognizes and responds to the scarcity, gradual destruction and aggravated pollution of water resources. In short, these new demands and uses must ultimately be a constraint on traditional, often inefficient, uses of water.

The objectives of Integrated Water Resources Management as authoritatively articulated in Agenda 21 (Chapter 18, article 18.9 (1992)) are:

a. To promote a dynamic, interactive, iterative and multisectoral approach to water resources management, including the identification and protection of potential sources of freshwater supply, that integrates technological, socioeconomic, environmental and human health considerations;
b. To plan for the sustainable and rational utilization, protection, conservation and management of water resources based on community needs and priorities within the framework of national economic development policy;
c. To design, implement and evaluate projects and programmes that are both economically efficient and socially appropriate within clearly defined strategies, based on an approach of full public participation, including that of women, youth, indigenous people and local communities in water management policy-making and decision-making;
d. To identify and strengthen or develop, as required, in particular in developing countries, the appropriate institutional, legal and financial mechanisms to ensure that water policy and its implementation are a catalyst for sustainable social progress and economic growth.

Between the Earth Summit in 1992 and the World Summit on Sustainable Development in 2002, IWRM was endorsed by the Commission on Sustainable Development, the General Assembly of the United Nations, and the Ministerial Declaration of the International Conference on Freshwater.[5] WSSD, as we have seen has not only endorsed it but has called for the development of IWRM plans by 2005.[6]

This is a daunting task and one that calls for a better understanding of the multiple facets of water resources management. This chapter explains the implications of an ecological approach which constitutes one facet of IWRM, but makes only a preliminary and rudimentary attempt to clarify the political, economic and social facets of IWRM.

From Theory to Practice

A comprehensive ecological approach confronts a substantial difficulty that needs to be addressed at the outset. If everything is related to everything else in increasing degrees of complexity, then nothing can be done unless everything is understood. An integrated approach seeks to further such understanding by synthesizing the myriad areas of knowledge about these ecologically interconnected issues. But, fully understanding the daunting biocomplexity of the physical, biological, and chemical life support systems in order to undertake IWRM is an intimidating, if not an impossible, task. An integrated approach seems first to demand an almost superhuman feat of comprehending all these earth support systems. This is followed by the need for legal and institutional implementation, although the legal and administrative structures required for such an endeavor are non-existent today.

The eminent economist/political scientist Charles Lindblom articulated the significant deficiencies of an integrated approach; he cogently argued that precisely because everything is interconnected, environmental problems are beyond our capacity to control in one unified policy (Lindblom, 1959). Lindblom asserted that the very enormity of the interconnected environment makes it impossible to treat it as a whole. Tactically defensible or strategically defensive points of intervention must be found (Lindblom, 1973), suggesting that a step-by-step approach will help to solve a problem better than a grand solution based upon the necessarily incomplete analysis offered by comprehensive rationality.

Lindblom also contended that a 'rational-comprehensive' decision-making process that adopts a synoptic perception of a problem, collects all relevant information, and explores all relevant solutions after considering all relevant answers, is impossible to develop when dealing with the environment as a whole. Such an approach, which is admirably marked by clarity of objective, explicitness of evaluation, a high degree of comprehensiveness of overview, and quantification of values for mathematical analysis, is only possible when dealing with small-scale problems with a very limited number of variables.

Lindblom suggested that poor as it is, incremental politics ordinarily offers the best chance of producing beneficial political changes.

To the extent that IWRM must take account of political reality, it seems undeniable that most people simply find it too overwhelming to think concurrently of whole litanies of problems without succumbing to agitated confusion or passive despair. Instead, building a series of 'small wins' creates a sense of control, reduces frustration and anxiety, and fosters continued enthusiasm on the part of the public, scientists, and politicians (Heinen and Low, 1991). These 'small wins', however, can be real victories only if they contribute to an overarching integrated water resources management strategy.

Lindblom notwithstanding, ecologicalism calls for the adoption of a comprehensive approach to the entire planetary ecosystem, and to its management. This is a gargantuan scientific task. The need to better understand water, the atmosphere, and biodiversity demands rigorous scientific research not only into whole systems, but also into how the details fit into the overall scheme (Trudgill, 1988). This involves not only large-scale control experimentation on whole catchments and ecosystems, but also the modeling of whole systems in order to study the likely consequences of management actions.

However, management relating to ecosystems or risk, whether on a national or international level, is a largely political activity in which the decisions taken may not be based on the best scientific evidence (Ruckelshaus, 1997; Trudgill, 1988). That is why it is important for at least some scientists to be conscious of the political dimension to their research, so that it might be presented in the most politically palatable fashion, and in a manner that lends itself to legal and administrative adoption.

The difficulties are compounded by the fact that Sustainable Development and Integrated Water Resources Management must take account of quintessentially political concepts and constructs such as economic and social development. If IWRM is to arise from the interconnected maze of ecological, social and economic factors, it becomes even more important to demonstrate how this might be achieved. Two possible examples are offered by Habitat Conservation Plans (HCP)[7] under the Endangered Species Act, and Integrated Watershed Management under the Clean Water Act (CWA). While HCPs have the potential to become instruments of IWRM at large geographical scales, they have not explicitly been used in this fashion (Ebin, 1997). IWM efforts, on the other hand, possess a historical, political and legal lineage that locates them within the IWRM family, and it is to IWM that we now turn.

4.   INTEGRATED WATERSHED MANAGEMENT: HISTORICAL BACKDROP

The 21st century concept of Integrated Watershed Management in the United States is based upon the evolving concept of Unified River Basin Management (URBM). The water resources management community has engaged this familiar, albeit changing concept, from around 1900 (Allee et al., 1982; Wengert, 1981). During this century-long period, URBM has come full circle. It began by emphasizing unified multipurpose river basin management of large systems. Subsequently it was reconfigured to deal with regional developmental projects, eventually expanded to include non-structural, basinwide flood and pollution control. In the past three decades it has been displaced by technology-based 'rule of law' pollution standards as the 'big dam' era wound down. More recently, it is being restored to a more comprehensive version of Unified River Basin Management called Integrated Watershed Management. Two features of URBM are worthy of initial note.

First, URBM is based on divergent streams of policy and law that roughly represent the 'sustainable and "developmental' facets of Sustainable Development (see, for example, Allee et al., 1982; Doll, 1994; Holmes, 1972, 1979; Schad, 1989). The first stream of policy and law envisioned watershed management as a comprehensive undertaking reflecting ecological principles. This comprehensive concept may be identified with the 'sustainable' part of SD. The second stream, setting up specific developmental projects consisting of single-purpose or 'multi-objective' statutes, reflects the 'development' face of SD.

Second, URBM, at least in the first half of the 20th century, was an unmistakably utilitarian or instrumental 'conservationist' construct, as distinct from the 'preservationist' environmental concept.[8] Consistent with the philosophy of the times, URBM was not aimed at the ecological or environmental protection of watersheds per se, but at the 'comprehensive development of river basins for multiple purpose use of water resources' (Schad, 1989, pp. 9–10). The purposes referred to were utilitarian in nature and dealt with navigation, irrigation, flood control, and hydropower (Doll, 1994; Wengert, 1981). The manner in which URBM emerged contains lessons for IWM, and it is to this historical development that we briefly turn.

The comprehensive approach to western water management originated in the suggestions of John Wesley Powell (Powell, 1879; Worster, 2001). The idea of integrated river basin water policy was developed during the Progressive Era in a series of reports issued by various commissions under Theodore Roosevelt's Administration: the 1908 Inland Waterways Commission, the 1909 National Conservation Commission, the 1912 National Waterways Commission, and the authorized but never formed 1917 Newlands Commission (Doll, 1994; Schad, 1989; Wengert, 1981).

The grand Progressive Era river basin proposals were never adopted by Congress, for reasons ranging from inattention, to a lack of willingness to cede congressional authority to the Executive, to the development-oriented sources of support (Adler, 1995). Instead, individual federal agencies, including the Bureau of Reclamation, the United States Army Corps of Engineers, and the Federal Power Commission proceeded with the construction and licensing of mission-oriented water development projects which were often not, in fact, consistent with the earlier vision of a comprehensive river basin plan.

However, multipurpose, basinwide water resources development, albeit dealing primarily with hydrologic and engineering problems, was advanced substantially by the Reclamation act of 1902,[9] waterways commission reports of 1908–17, the Federal Power Act of 1920,[10] the Corps of Engineers '308' basin study authorizations of 1925–27, and the Boulder Canyon Project Act of 1928[11] (Wengert, 1981). The underlying principle of multipurpose, basinwide water resources development was that large, federally planned and funded impoundments could stimulate basinwide economic development by combining flood control, municipal water supply, irrigation, hydroelectric power generation, recreation, and water quality improvement functions within single projects.

A broader view of comprehensive river basin planning as an economic development construct was ushered in by the New Deal in proposals by the National Planning Board, the Water Resources Committee of the National Resources Commission, and the National Resources Planning Board (Schad, 1989). As with Progressive Era proposals, the New Deal agencies suggested a 'comprehensive approach integrating all resources into a unified, balanced program' (Wengert, 1981), or in the words of President Franklin D. Roosevelt, 'a thoroughly democratic process of national planning for the conservation and utilization of the water and related land resources of our country' (Dworsky and Allee, 1981, p. 38, quoting Roosevelt's Message to Congress, 3 June 1937). These proposals resembled today's watershed proposals somewhat more closely than Progressive Era versions with their increased recognition of the link between land development (including deforestation) and water resources degradation caused by increased erosion and run-off.

Like the Progressive Era proposals before them, however, the New Deal watershed proposals were fundamentally rooted in human use of water and economic development. In addition to traditional outputs such as flood control, navigation, irrigation, and hydropower, and consistent with New Deal programs as a whole, these proposals sought to use comprehensive river basin planning and development to provide jobs and promote regional economic development (Doll, 1994; Wengert, 1981).

A new phase in water management policy should have begun with the Water Resources Planning Act of 1965[12] which established a federal interagency Water Resources Council to supervise and implement comprehensive, coordinated joint plans (CCJPs) prepared by river basin commissions consisting of federal, state, and local officials.[13] The CCJPs were to be followed by more detailed basin studies (level B studies), which, in turn, were to form the bases for water resources development project planning (Allee et al., 1982). This detailed planning mechanism was, in general, a failure, although a number of valuable CCJPs were produced. A significant reason for this failure lay in the emergence of a new stream of environmental laws and policies beginning in the late 1960s and culminating in the 'environmental decade' of the 1970s that were largely driven by what one of us has called the '"rule of law" ideology' (Tarlock, 2000).

The 'rule of law' ideology helped to bring about the triumph of environmentalism over developmentalism. It also signaled the triumph of 'command and control' environmental regulation over more flexible concepts based on subjective expertise. The crucible of environmental ideas in the 1960s gave rise to two different currents of thinking. On the one hand, environmentalism in the late 1960s was rooted in holistic and ecological thinking which found expression in the enactment of the National Environmental Policy Act (1969) and the creation of the Environmental Protection Agency (EPA). On the other hand, serious doubts about whether the New Deal belief in independent and expert administrative agencies could creatively regulate a complex social problem in the public interest affected the approaches taken to environmental problems.

As we have noted beliefs in interconnected ecosystems offer a holistic, rather than a fragmented, view of the world. Translating this worldview into practice requires the integration of political and administrative policies dealing with the environment. Air, water and land were part of one environment and did not constitute separate and discrete entities. However, the complex and uncertain nature of environmental problems did not admit of preordained solutions. To legislate in advance on how the balance should be struck in the myriad of situations crying out for solutions would only create procrustean beds. Pollution control required fine and expert balancing that could best be done by expert and sensitive agencies vested with power over the whole environment and empowered to act in the particular circumstances of the case. An integrated approach called for a broad delegation of power. Arguments for integration based on ecological thinking, however, were countered by others that resisted the granting of wide discretionary power.

During the New Deal, champions of the administrative process prevailed with their view that there was an objective public interest that could be ascertained and implemented by expert administrators (Stewart, 1975; Sunstein,

1987). Their approach came under heavy attack from political scientists on constitutional and political grounds (Jaffe, 1956). The constitutional objections have only recently been resolved in favor of expert administrators (Breyer and Stewart, 1985).[14] The political arguments have been more successful. Those attacking the technocratic philosophy charged that independent agencies, having no duly constituted master, were falling under the domination of private interests, usually the very interests whose activities they were supposed to regulate (Fellmeth, 1970; Turner, 1970). A somewhat different criticism was leveled by economists who saw regulation as being inefficient because it was created and administered for the benefit of well-organized interests at the expense of the public. These critics either advocated deregulation or regulatory reform (Breyer and Stewart, 1985). The economic arguments take the form of two interrelated theories: the doctrinaire theorists call for deregulation, the abolition of agencies, and a return to markets based on the assumption that no regulatory process can ever be responsive enough to replicate the efficiency of the market, and second that, in any event, efficient regulation is impossible because regulatory agencies are colonized by those who pursue their self-interest (Stigler, 1971; Stigler and Friedland, 1962). The more pragmatic reformers, meanwhile, argue that regulation is a necessary 'market-supporting' mechanism for dealing with market failures associated with environmental externalities (Breyer, 1982; Winter, 1973). Ironically, political and economic critics of regulation agreed that regulation benefited the regulated rather than the public.

By the end of the 1960s, much of the regulation in the United States was seen to be in 'deep trouble' (Noll, 1971, p. 110). It became necessary to face up to the problem of how agencies had misused and even abused the broad delegated power conferred upon them. Confidence in the ability of administrative agencies to implement statutes effectively and in the public interest had apparently evaporated. Many influential commentators referred to the problems arising out of the unsatisfactory or inadequate implementation of the legislative mandates given to administrative agencies. They suggested that one way of remedying this problem lay in statutes with clear mandates and definite obligations. One reason for the malaise was the nature of the legislative mandate; presumably, statutory mandates lacked clarity and rarely provided clear directions to the new agency (Bernstein, 1955).[15] The vagueness was deliberate and resulted from the lobbying of well-organized private groups who were the subject of the regulation. Having failed in their efforts to prevent the enactment of legislation affecting them, these private groups concentrated on making the regulatory provisions as vague and innocuous as possible, confident that they could 'capture' the agency in question. The unwillingness or inability of Congress to give better directives to its agencies was also criticized (Friendly, 1962).

Theodore Lowi, and others, have synthesized the criticisms of the New Deal agencies and suggested that one remedy for many of their troubles might lie in statutes that have clear goals and explicit means of implementation. Others have demanded that the agencies should redeem their New Deal promise by generating clear standards through creative rule making. Another solution is to look to the courts for action.

These new statutory, administrative and court-ordered norms would target and institutionalize the public needs which led to the statute in the first place, and would make it difficult for the agency to postpone the performance of its obligations. One of the central themes present when environmental legislation was being formed, therefore, was that expertise could be an excuse for inaction, and even worse, could be captured by special interests. The remedy suggested by believers in regulation was the enactment of legislation setting forth explicit goals, specific means by which these goals could be attained, and rigorous timetables in which to do so.

Pursuant to this thinking, new environmental standards were adopted in the Clean Water Act (CWA) of 1972. Prior to 1972, the Federal Water Pollution Control Act had established a state-administered system based on ambient water quality standards. While these ambient quality standards did allow for a more comprehensive approach, the very restrictive conditions placed on enforcement, coupled with the inability or disinclination of states to enforce these water quality standards led to a failure of implementation, and the deterioration of water quality. Responding to this challenge in 1972 and 1977, Congress adopted the environmental thinking of the time and enacted a totally different approach to pollution control and water quality based on identified problems and their control through permits based on a variety of technology-based standards.

Meanwhile environmentalist reassessments of Unified River Basin Management stressed non-structural, basinwide flood and pollution controls, including wetlands preservation, critical area protection, and restrictions on floodplain development (Allee et al., 1982; Wengert, 1981). Environmentalists also called attention to the deleterious environmental impacts of large-scale irrigation, hydroelectric water supply, and water-based recreation projects. The antidevelopment orientation of environmentalists was frequently shared by fiscal conservatives, because 'most of the major and more dramatic water projects were under construction or had already been completed [and] ... [t]hose that remained were, by definition, small projects and more doubtful from a [benefit/cost] point of view' (Wengert, 1981, p. 24).

Environmentalists and fiscal conservatives also tended to agree that future water resources development projects should be primarily funded by user charges, not federal grants. Acting on his fiscal conservatism, President Ronald Reagan terminated the Water Resources Council and the river basin commissions in 1981, although the Water Resources Planning Act has not formally been repealed (Struck, 1982). This period came to an end with the completion or deauthorization of most of the major federal water resources development projects previously authorized.

A necessary corollary of the effort to identify specific problems was that the Clean Water Act approached complex water problems by fragmentation, 'attempting to carve the larger puzzle into smaller pieces that can be isolated and micromanaged through categorical command style rules' (Karkkainen, 2002, p. 555). While the National Pollution Discharge Elimination System (NPDES) established by the 1972 Federal Water Pollution Control Act Amendments did substantially reduce point source discharges, it was unable to control non-point sources, and large extents of water remain unacceptably dirty (Houck, 1999). Moreover, the attempt to control water pollution independent of air and land-based pollution ignored the effect of cross-media pollution. For example, air pollution (atmospheric deposition) accounts for 25 per cent or more of nitrogen pollution in water bodies such as Chesapeake Bay (Houck, 1999). Such fragmented piecemeal regulation is ineffective because it is crafted on the basis of insufficient and incomplete information. It is also inefficient because it imposes redundant costs on both regulators and regulated (for example, see Guruswamy, 1989; Karkkainen, 2002; Rondinelli, 2001; Stewart, 2001).

Mindful of these facts, the Clinton Administration acknowledged the limitations of top–down rule making and enforcement as a method of solving complex environmental problems. That administration began the task of 'making the transition from a clean water program based primarily on technology-based controls to water quality-based controls implemented on a watershed basis' largely inspired and based around the TMDL program (Houck, 1999 [citing a Memorandum from Robert Perciasepe, Assistant Administrator for Water, United States Environmental Protection Agency (8 August 1997)]). The TMDL (total maximum daily load) program is a tool outlined in section 303(d) of the Clean Water Act, and applies to streams that do not meet water quality standards established by the states under Section 303. TMDL programs calculate allowable discharges based on the cumulative assimilative capacity of a given stream, rather than imposing uniform point source technology standards not directly tied to stream conditions (Houck, 1999).

The Clinton Administration made significant, if ad hoc, steps toward making integrated, collaborative, regional ecosystem and watershed management an important aspect of environmental and natural resources policy (Babbitt, 2000; Frampton, 1996; Karkkainen, 2002). This is evident in places like the Everglades,[16] the Chesapeake Bay,[17] the San Francisco Bay-Delta,[18] river basins of the Pacific Northwest,[19] and critical watersheds more generally.[20]

The Bush Administration has attempted to move the integration strategy to a new level by transforming such regional efforts into a national strategy, and has placed even more formal emphasis on a holistic approach to water quality management. In launching a new Watershed Initiative for the entire country, the EPA has stated that its comprehensive approach to the health of aquatic resources 'recognizes needs for water supply, water quality, flood control, navigation, hydropower generation, fisheries, habitat protection, and recreation – and it recognizes that these needs often compete'.[21] Its will to implement such a strategy remains very much in doubt.

5. THE LEGAL CHALLENGE

Jurisdiction

The legal basis for such a comprehensive approach is somewhat tenuous. American political boundaries do not, for the most part, correspond to water basins. Thus, there is rarely a single competent institution with legal jurisdiction over a water basin. In addition, water resources problems and activities, such as interbasin transfers of water, often transcend even recognized regional boundaries. This institutional situation creates the traditional incentive for one jurisdiction to solve its own development problems without regard to spill-over water resources effects on neighboring jurisdictions, frequently downstream or down-gradient jurisdictions (Ingram, 1973). Moreover, regional solutions to water resources management problems are also frustrated by the difficulty of defining a water basin in a way that will both promote holistic problem solving and elicit political support.

As we have seen, however, important facets of a comprehensive approach have functionally been implemented in river basins, and the successful accommodation of key goals of integration within existing frameworks of laws and policies provides momentum for their wider application. As we have noted, key EPA personnel see Section 303(d) of the Clean Water Act (CWA) as offering a legal basis for Integrated Watershed Management. Other commentators, albeit with some reservations, have suggested that a legal basis for integrated water basin management may be found within Section 303(d) read with Section 401 of the CWA (discussed later) (Houck, 1999; Ruggiero, 1999).

The draft total maximum daily load (TMDL) programs drawn up by states are subject to review and approval by EPA. To the extent that they are ambient as opposed to end of pipeline standards, the water quality standards under the CWA are comparable to national ambient air quality standards set for criteria pollutants under the Clean Air Act (CAA). However, the CWA standards lack the obligatory and mandatory characteristics of their CAA counterparts.

One of the possible objections to a TMDL approach to IWM is that TMDLs are essentially pollution control mechanisms. They may be so but water quality standards under the CWA can establish criteria for standards based on designated uses, maintaining minimum flows, maintaining certain temperature regimes, protecting aesthetic values, maintaining natural plant and animal assemblages, and more. Almost all of these criteria are linked to events and conditions outside the water body itself; and these criteria are closely linked to the condition of the watershed/ecosystem that a water body drains. To this extent, TMDLs are not focused only on the pollutants in a water body. To the contrary, section 303(d) requires land managers to protect water quality where it interfaces with terrestrial ecology, and thereby ties water quality to the overall condition of watersheds. Thus 303(d) does facilitate IWM, because TMDLs based on ambient water quality standards permit the regulators and the regulated to adopt a more comprehensive approach that takes account of all sources of water pollution.[22]

According to some commentators the TMDL program has never seemed farther from implementation. They point out that state governments have shied away from their environmental responsibilities under the TMDL program, while the Bush Administration has withdrawn the final regulatory program. What this demonstrates is a lack of political will to implement TMDLs, at both the state and federal levels (Malone, 2002; Houck, 2002). In addition to political resolve, IWM calls for an integration of water pollution with air pollution and this objective will face formidable bureaucratic objections within the EPA. While these reservations and challenges may well be valid, political obstacles and programmatic objections do not constitute jurisdictional challenges, and at this point we argue only that TMDLs provide a jurisdictional canopy under which IWM can be implemented. The jurisdiction to act is a necessary precondition to IWM, but such legal authority does not bestow, or even presage, political will or ecological management expertise.

Section 401 of the CWA is a licensing provision. It stipulates that before a federal permit or license may be granted for any activity which might result in a discharge into the nation's waters, the applicant must first obtain a state water quality certification. A water quality certification is essentially a state permit which says that the anticipated activity complies with the applicable effluent limitations, water quality standards or 'any other appropriate' state law requirements. When a state issues a certification, any standards or limitations contained in the certification become conditions of the federal license or permit. Since 401 permits are directed at ensuring compliance with state water quality standards they thereby enjoy a parity of status with TMDL's as instruments of IWM.

Sections 303(d) and 401 of the CWA, however, were not designed to create an integrated system of watershed management, and the final report of the National Watershed Commission convened by the United States EPA recognizes that it may become necessary to change or amend the CWA. But this may not be necessary, and it appears that IWM is proceeding as envisaged, without being paralyzed by the fear of jurisdictional challenge.

While the legal and political contours of IWM are being finalized in the United States, it is important to underscore the extent to which IWM is consistent with, and indeed constitutes a felicitous way of implementing Integrated Water Resources Management (IWRM). The United States EPA has laid out the manner that this might be done, but any attempt to reconfigure water management through a comprehensive approach brings up the question of water rights or entitlements.

Rights or Entitlements

The approach suggested in this chapter will ultimately require that water use decisions be measured by both traditional standards of available and beneficial use as well as Sustainable Development. This idea is not new in western water law. Water use has always been limited by the beneficial use doctrine and rights have been subordinated to public interest standards and the public trust doctrine. However, the primary function of water law has been to create secure rights with minimal limitations on use and enjoyment (Hobbs, 2002). Integrated Watershed Management will change this expectation because it requires that decisions be made at larger regional scales, and thus it will add a significant new public interest component to existing water rights. The new decision-making institutions that follow will never displace the core water allocation institution – the law of water rights – but they will serve to put a floor under these uses to balance the maintenance of aquatic ecosystem integrity with traditional consumptive uses.

Water use entitlements must be re-examined because they are central to any sustainability policy. There is a widespread agreement that many uses – for example, groundwater mining, excessive irrigation from crops to golf courses and wasteful urban uses – are not sustainable. Environmentalists have identified the existing structure of property rights as a barrier to sustainability (Freyfogle, 1998). It is hard to talk of restoring a river if the rights to use the river are fully assigned to consumptive uses. In return, western water users have argued that the Constitution permits only two types of modification: (1) voluntary transfers of an existing right or (2) compensated, involuntary transfer. We argue that both views are both right and wrong, and that it is possible to modify existing rights to promote greater sustainability consistent with basic principles of fairness.

We prefer the term 'entitlements' to 'rights' because we believe that it gets closer to the right question: that is, what the reasonable expectations of all users should be, compared to the more freighted term 'rights'. We argue that sustainable development requires both the presence of secure entitlements to use water and the modification of existing entitlements. Entitlements have two major advantages over the pure administrative assignments of the privilege to use water. First, entitlements provide the security needed to invest in the necessary water infrastructure that can be a new storage facility or the purchase of existing rights to dedicate to instream flow maintenance. Second, entitlements are necessary to trigger the emergence of markets that are an important (but not exclusive) component of any reallocation strategy. Reallocation in turn is a necessary element of any move towards sustainable development. Entitlements carry with them the duty of responsible or sustainable use.

Environmentalists are correct to identify our concept of entitlements to use resources as the root of the problem, but have overexaggerated the lack of flexibility in existing water 'rights' (Rose, 2000).[23] As many students of environmental regulation have forcefully observed, environmental regulation is a modest overlay and modification of the liberal institutions of private property and consumer sovereignty (Westbrook, 1994). Regulation does not challenge the fundamental idea that individuals may determine the amount of resource consumption subject only to the caveat that they internalize some portion of the more obvious social costs of consumption. The external cost minimization justification for environmental regulation contemplates the possibility of resource use reduction as the gap between private and social cost is narrowed, but we do not think that there is much evidence of serious changes in consumption patterns. In contrast, Sustainable Development seeks to identify the root causes of behaviors that cause environmental degradation and to change them substantially. It seeks to modify rather than mitigate.

Integrated Watershed Management will require the more intense management of water basins and will inevitably change the nature of water entitlements. This is already taking place in a number of 'out of box' ad hoc basin management experiments in the West including the Columbia, Sacramento-San Joaquin, Snake, Platte, and Rio Grande rivers. The experiments try to work within existing water rights, but they are in fact modifying the concept of entitlement.

The case for a new conception of water use entitlements is both empirical and normative. The empirical case is based on the fact that the West faces changed conditions that require a careful reconsideration of the underpinnings of the region's traditional water allocation conditions.[24] The concept of sustainable water entitlements builds on two critical aspects of water rights, which have always differentiated them from land rights.

The first is that water entitlements are inherently incomplete property rights.[25] They are incomplete for two reasons. First, they are, of necessity, correlative. The use of the resource must be shared with other similarly situated users. Sharing among stream owners is the heart of the common law of riparian rights, but it also exists in the law of prior appropriation. The beneficial use doctrine, the rule that unused water returns to the system and is open to reappropriation, and the protection of junior right holders in transfers, are all sharing rules. Second, water entitlements carry inherent risks. The traditional risk, of course, is short-term climate variability. Our drought policy has always been twofold: build carry-over storage and hope for rain. Entitlement holders have learned to live with these risks, and, understandably, argue that 'traditional' climate variability marks off the limit of risk that an entitlement holder assumes. We argue, however, that there is a need to recognize that water use entitlements are not static. The classes of uses with which the entitlement must be shared and the risks to which the entitlement holder is subject are subject to expansion. This is not an argument that existing entitlements can be reallocated by courts, legislatures and administrative agencies without regard for historic uses. It is an argument that the essence of a use entitlement is not necessarily the formal definition of the right but the amount of water reasonably necessary to sustain the water-dependent activity. This recognition is an integral first step to encourage entitlement holders to consider voluntary integrated management options that present acceptable, bounded modifications of their current risk levels.

CONCLUSION

The competing demands for water have focused national and international attention on the need for a more comprehensive approach to water management. The fact that Integrated Water Resources Management is emerging as a norm of international environmental law places some obligation on the United States to act consistently with such an evolving concept. The American West offers promising illustrations of the potential and the challenges confronting such a comprehensive approach. It is encouraging that the United States is in fact moving toward a similar objective through Integrated Watershed Management. The extent to which this concept will grow to fruition depends on the clarification of the legal foundations particularly as they pertain to administrative integration and to property rights and entitlements. We are witnessing the beginnings of a legal and political discourse about this unfolding saga.

NOTES

1. The current focus is on restoration because most river systems have been modified. The best we can do is recognize that rivers perform valuable functions from the maintenance of consumptive uses to ecosystem services, and try to manage them to deliver an acceptable level of these services.

2. A groundnorm, translated in the US as the basic norm, is the foundational premise or initial hypothesis conferring validity or legitimacy on all other norms (Kelsen, 1967).

3. The United Nations Conference on Environment and Development (UNCED) or Earth Summit was convened in Rio de Janeiro, Brazil, in 1992. The Earth Summit was the biggest and most important environmental conference in history. It sought to give expression to sustainable development, and fulfill its goals of addressing the dual problems of environmental protection and socioeconomic development, by producing two treaties: the Convention on Biological Diversity, and the Framework Convention on Climate Change; two instruments: the Rio Declaration, and Agenda 21; together with a non-binding declaration on Forest Principles.

4. As defined in *Encyclopedia of Life Support Systems* (www.eolss.co.uk): 'A life support system is any natural or human-engineered system that furthers the life of the biosphere in a sustainable fashion. The fundamental attribute of life support systems is that together they provide all of the sustainable needs required for continuance of life. These needs go far beyond biological requirements. Thus life support systems encompass natural environmental systems as well as ancillary social systems required to foster societal harmony, safety, nutrition, medical care, economic standards, and the development of new technology. The one common thread in all of these systems is that they operate in partnership with the conservation of global natural resources.'

5. *See* the 'Report of the Expert Group Meeting on Strategic Approaches to Freshwater', Paragraph 11 (1998) www.un.org/documents/ecosoc/cn17/1998/background/ecn171998-freshrep.htm; the 'International Year of Freshwater – UN General Assembly Resolution A/RES/55/196', 1 Feb 2001, www.unesco.org/water/iyfw/res_55_196e.pdf; and the 'Ministerial Declaration adopted by Ministers Meeting in the Ministerial Session of the International Conference on Freshwater', Bonn, 4 December 2001, www.water-2001.de/outcome/Ministerial_declaration.asp.

6. For details, see the WSSD 'Plan of Implementation', Article 25 (2002).

7. Habitat Conservation Plans (HCPs) provide a framework within which private and public parties can establish an integrated and regionally coherent strategy for endangered species recovery that, to the extent possible, is coordinated with existing and proposed human activities. HCPs are designed not only to protect species, but also provide some protections to private landowners from liability under the Endangered Species Act.

8. Conservationists view the environment as the biophysical life support system of humans, and the material basis for economic activity. According to these instrumentalist and utilitarian analysts and thinkers, to act rightly toward the earth is to allocate resources efficiently. The point of environmental policy and law would not be to protect nature for its own sake but to maximize the long-run benefits nature offers humankind. In the US this view was articulated by the 'conservationists' represented by Gifford Pinchot, and is kept alive today by the ever-expanding domain of economic analysis. In contrast, the 'preservationists' argue for treating the earth with reverence and respect and thus not violating its beauty or integrity. They value nature not simply because of the good it does us but also for its own sake because it is good in itself. In the US, this seam of thinking has perhaps been best articulated by John Muir and Frederick Olmstead.

9. Act of 17 June 1902, ch. 1093, 32 Stat. 388 (1902).

10. Federal Power Act, 41 Stat. 24 (1920) (codified as amended at 16 U.S.C. § 791a et seq. (1993)).

11. Boulder Canyon Project Act, ch. 42, 45 Stat. 1057 (1928) (codified as amended at 43 U.S.C. § 617 et seq. (1992)).

12. 42 U.S.C. §§ 1962–1962d–20 (1988 & Supp. III 1991).

176   *In search of sustainable water management*

13. 42 U.S.C. § 1962a (1988).
14. The constitutional objections were threefold: first, that all executive functions should be subject to Presidential control (and that independent agencies were a headless fourth branch of government); second, that these independent agencies combined powers previously distributed among the three traditional branches; and third, the non-delegation doctrine was briefly revived (unsuccessfully) to argue that the powers of independent agencies were inadequately defined and constrained. (For additional details, see Breyer and Stewart (1985) and *Whitman v. Amercian Trucking Associations*, 531 U.S. 457 (2001).)
15. Bernstein set up an influential model of agency obsolescence in which he traced the cycle of a regulatory agency from gestation to youth, youth to maturity, and maturity to old age when the agency suffered debility and decline and 'surrendered' to the regulated.
16. The South Florida ecosystem restoration project is a joint effort of the Florida Department of Environmental Protection, the South Florida Water Management District, the Army Corps of Engineers, the Department of the Interior, the EPA, other state and federal agencies, and private and nonprofit participants (Ansson, 2000; Light et al., 1995). It seeks to restore ecosystem health in the Everglades and associated ecosystems while also ensuring adequate public drinking water supplies and flood control by re-engineering water diversions, reducing non-point source pollution, restoring degraded habitats, and managing the hydrology of this region, where water flows are the principal defining characteristic of the landscape.
17. Although the Chesapeake Bay Program has been in existence since the early 1980s and its historical roots predate the EPA itself, it has gained new prominence as a model of integrated watershed management (Cannon, 2000; Costanza and Greer, 1995).
18. The Bay-Delta program is an ambitious collaborative effort by federal and state agencies aimed at integrated management and ecological restoration of the San Francisco Bay/Sacramento–San Joaquin Delta estuary system (Wright, 2001). The Sacramento and San Joaquin Rivers are the principal tributaries of the San Francisco Bay. They also supply drinking water to some 20 million water-starved Californians as far away as San Diego, as well as irrigation water for the rich agricultural region of the Central Valley. State–federal cooperation was formalized in 1994 with the signing of a Framework Agreement creating CALFED, a hybrid state–federal institution involving the Department of the Interior, the EPA, the Department of Agriculture, the National Marine Fisheries Service, the Army Corps of Engineers, and corresponding state agencies.
19. Regional water management in the Pacific Northwest has largely been inspired by the demands of recovering endangered salmon (Blumm, 1997; Conway and Evans, 2000; Volkman, 1999).
20. A variety of examples exist, from the recommendations of the Committee of Scientists (1999) using integrated ecosystem management and collaborative planning as the foundation of national forest and grassland management in the twenty-first century, from the watershed restoration programs spearheaded by EPA (1996).
21. www.epa.gov/fedrgstr/EPA-WATER/2002/August/Day-20/w21196.htm
22. Note that in *Proscolino v. Nastri*, 291 F.3d 1123 (9th Cir. 2002), the court held that TMDLs may include non-point sources.
23. Property has been a dynamic and changing concept, as Native Americans well know, in the United States.
24. This argument draws on the progressive tradition of jurisprudence. Harr and Wolf (2002) identify five questions as characteristic of progressive era jurisprudence: (1) does the challenged regulation reflect the elasticity and adaptability of traditional common law methodology? (2) was the challenged regulation crafted by important input by experts from nonlegal fields? (3) does the challenged regulation have the capacity to reduce, and at the same time, enhance individual wealth and personal rights? (4) is the Court being asked to affirm judicial and popular acceptance in the 'laboratory' of states? and (5) is the regulation fundamentally flexible, in that it furthers a wide range of public interests and features exemption provisions?
25. This characterization borrows from two intellectual constructs. Modern law and economics theory distinguishes between entitlements protected by property rule (injunctive relief)

and those protected by only liability rules (damages) (Calabrisi and Melanred, 1992). Public utility law distinguishes between firm and interruptible service. In general, property rules are recognized when a proposed allocation is presumed efficient and the transaction costs of a reallocation are low; damages are preferred when transaction costs are high (Posner, 1997). This analysis makes the crucial assumptions that a property owner's expectations remain constant over time and that we must protect the right in a consistent manner. It also pays insufficient attention to the possibility of mutual dynamic mitigation. In fact, more generally, all or nothing legal solutions are breaking down throughout the law; we can devise solutions that take advantage of the positive benefits of both property and liability rules. Property rules enable parties to create stable future resources use regimes (Rose, 1997), but liability rules can provide efficiency gains that property rules sometimes block.

LITERATURE CITED

Adler, Robert W. (1995), 'Addressing barriers to watershed protection', *Environmental Law*, **25**, 973.

Allee, David J., Leonard B. Dworsky and Ronald M. North (eds) (1982), 'United States water planning and management', in *Unified River Basin Management – Stage II*, Minneapolis, MN: American Water Resources Association, pp. 11–42.

Ansson, Richard J. Jr. (2000), 'Ecosystem management and our national parks: will ecosystem management become the guiding theory for our national parks in the 21st century?' *University of Baltimore Journal of Environmental Law*, **7**, 87.

Babbitt, Bruce (2000), 'Restoring our natural heritage', *Natural Resources and Environment*, **14** (Winter) (3).

Bates, Sarah F., David H. Getches, Lawrence J. Macdonnell and Charles F. Wilkinson (1993), *Searching Out the Headwaters: Change and Rediscovery in Western Water Policy*, Washington, DC: Island Press.

Bernstein, Marver H. (1955), *Regulating Business by Independent Commission*, Princeton, NJ: Princeton University Press.

Blumm, Michael C. (1997), 'The amphibious salmon: the evolution of ecosystem management in the Columbia River Basin', *Ecology Law Quarterly*, **24**, 653.

Botkin, Daniel B. (1990), *Discordant Harmonies*, New York: Oxford University Press.

Breyer, Stephen G. (1982), *Regulation and Its Reform*, Cambridge, MA: Harvard University Press.

Breyer, Stephen G. and Richard B. Stewart (1985), *Administrative Law and Regulatory Policy*, 2nd edn, Boston: Little Brown & Company.

Calabrisi, Guido and Douglas A. Melanred, (1992), 'Property rules, liability rules, and inalienability: one view of the cathedral', *Harvard Law Review*, **85**(6), 1089.

Cannon, Jon (2000), 'Choices and institutions in watershed management', *William and Mary Environmental Law Review*, **25**, 379.

Committee of Scientists, US Department of Agriculture (1999), *Sustaining the People's Lands: Recommendations for Stewardship of the National Forests and Grasslands into the Next Century*, Washington, DC: US Department of Agriculture.

Conway, Dianne K. and Daniel S. Evans (2000), 'Salmon on the brink: the imperative of integrating environmental standards and review on an ecosystem scale', *Seattle University Law Review*, **23**, 977.

Costanza, Robert and Jack Greer (1995), 'The Chesapeake Bay and its watershed: a model for sustainable ecosystem management?', in Robert Costanza and Jock Greer (eds), *Barriers and Bridges to the Renewal of Ecosystems and Institutions*, New York: Columbia University Press, p. 103.

Doll, Amy (1994), 'Evolution of watershed planning and management in national water policy', in *Watershed '93*, Washington, DC: US Bureau of Reclamation, pp. 107–13.

Dworsky, Leonard B. and David J. Allee (1981), 'Unified/integrated river basin management: evolution of organizational arrangements', in Ronald M. North et al. (eds), *Unified River Basin Management*, Minneapolis, MN: American Water Resources Association, p. 28.

Ebin, Marc C. (1997), 'Is the Southern California approach to conservation succeeding?', *Ecology Law Quarterly*, **24**, 695.

Environmental Protection Agency (EPA) (1996), *Watershed Approach Framework*, Washington, DC: Office of Water.

Farrier, David (2002), 'Protecting environmental values in water resources in Australia', paper prepared for the Natural Resources Law Center, University of Colorado School of Law, Boulder, CO.

Fellmeth, R. (1970), *The Interstate Commerce Commission: The Public Interest and the ICC*, New York: Grossman Publishers.

Frampton, George (1996), 'Ecosystem management in the Clinton Administration', *Duke Environmental Law and Policy Foundation*, **7**, 39.

Freyfogle, Eric (1998), *Bounded People, Boundless Lands*, Washington, DC: Island Press.

Friendly, H. (1962), *The Federal Administrative Agencies: The Need for Better Definition of Standards*, Cambridge, MA: Harvard University Press.

Guruswamy, Lakshman D. (1989), 'Integrating thoughtways: re-opening the environmental mind', *Wisconsin Law Review*, **1989**(2), 463.

Guruswamy, Lakshman and Jeffrey McNeely (eds) (1998), *Protection of Global Biodiversity Converging Strategies*, Durham, NC: Duke University Press.

Harr, Charles M. and Michael Allan Wolf (2002), 'Euclid lives: the survival of progressive jurisprudence', *Harvard Law Review*, **115**, 2158.

Heinen, J.T. and R.S. Low (1991), 'Human behavioral ecology and environmental conservation', *Environmental Conservation*, **19**(2), 105–16.

Hobbs, Gregory G., Jr. (2002), 'Priority: the most misunderstood stick in the bundle', *Environmental Law*, **32**, 37.

Holmes, Beatrice Hort (1972), *A History of Federal Water Resources Programs, 1800–1960*, publication no 1233,Washington, DC: US Department of Agriculture.

Holmes, Beatrice Hort (1979), *History of Federal Water Resources Programs and Policies, 1961–1970*, publication no 1379, Washington, DC: US Department of Agriculture.

Houck, Oliver A. (1999), *The Clean Water Act TMDL Program: Law, Policy, and Implementation*, Washington, DC: Environmental Law Institute.

Houck, Oliver (2002), 'The Clean Water Act TMDL Program V: aftershock and prelude', *Environmental Law Reporter*, **32**, 10385.

Ingram, Helen M. (1973), 'The political economy of regional water institutions', *American Journal of Agricultural Economics*, 55, 10.

Jaffe, Louis L. (1956), 'The New Deal Agency – a new scapegoat', *Yale Law Journal*, **65**, 1068.

Karkkainen, Bradley C. (2002), 'Environmental lawyering in the age of collaboration', *Wisconsin Law Review*, **2002**(2), 555.

Kelsen, Hans (1967), *Pure Theory of Law*, Berkeley: University of California Press.

Light, Stephen S. et al. (1995) 'The Everglades: evolution of management in a turbulent ecosystem', in Lance H. Gunderson, C.S. Holling and Stephen S. Light (eds), *Barriers and Bridges to the Renewal of Ecosystems and Institutions*, New York: Columbia University Press, p. 103.

Lindblom, Charles E. (1959), 'The science of muddling through', *Public Administration Review*, **19**, 79.

Lindblom, Charles E. (1973), 'Incrementalism and environmentalism', in *Managing the Environment*, Washington, DC: US Government Printing Office, p. 83.

Malone, Linda (2002), 'The myths and truths that threaten the TMDL Program', *Environmental Law Reporter*, **32**, 11133.

Matson, Pamela (2001), 'Environmental challenges for the twenty-first century: interacting challenges and integrative solutions', *Ecology Law Quarterly*, **27**, 1179.

Meyer, Judith L. (1993), 'Changing concepts of system management', in *Proceedings: Sustaining Our Water Resources*, proceedings of the Water Science and Technology Board Tenth Anniversary Symposium, Washington, DC: National Academy Press, p. 78.

Meyer, Judith L. (1994), 'The dance of nature: new concepts in ecology', *Chicago-Kent Law Review*, **69**, 847.

Noll, Roger G. (1971), *Reforming Regulation*, Washington, DC: Brookings Institution.

Noss, Reed E., and Allen Y. Cooperrider (1994), *Saving Nature's Legacy: Protecting and Restoring Biodiversity*, Washington, DC: Island Press.

Posner, Richard A. (1997), *Economic Analysis of Law*, 2nd edn, Boston: Little Brown & Company.

Powell, John Wesley (1879), *Report on the Lands of the Arid Region of the United States*, 2nd edn, Washington, DC: Government Printing Office.

Rondinelli, Dennis A. (2001), 'A new generation of environmental policy: government/business collaboration in environmental management', *Environmental Law Reporter*, **31**, 10891.

Rose, Carol M. (1997), 'Property rules, liability rules, and inalienability: a twenty-five year retrospective: the shadow of the cathedral', *Yale Law Journal*, **106**, 2175.

Rose, Carol M. (2000), 'Property and expropriation: themes and variations in American law', *Utah Law Review*, **1**, 1–38.

Ruckelshaus, William D. (1997), 'Risk, science and democracy', in Richard L. Revesz (ed.), *Foundations of Environmental Law and Policy*, New York: Oxford University Press, pp. 48–52.

Ruggiero, Jory (1999), 'Toward a law of the land: The Clean Water Act as a federal mandate for the implementation of an ecosystem approach to land management', *Public Land and Resources Law Review*, **20**, 31.

Schad, Theodore M. (1989), 'Past, present, and future of water resources management in the United States', in A. Ivan Johnson and Warren Viessman (eds), *Water Management in the 21st Century*, Middleburg, VA: American Water Resources Association.

Stein, Robyn (2002), 'Water law in a democratic South Africa: a country case study examining the introduction of a public rights system', paper prepared for the Natural Resources Law Center, University of Colorado School of Law, Boulder, CO.

Stewart, Richard B. (2001), 'A new generation of environmental regulation?', *Capital University Law Review*, **29**, 21.

Stewart, Richard B. (1975), 'The reformation of American administrative law', *Harvard Law Review*, **88**, 1667.

Stigler, George (1971), 'The theory of economic regulation', *Bell Journal of Economic and Management Sciences*, **2**, 3.

Stigler, George and Claire Friedland (1962), 'What van regulators regulate? The case of electricity', *Joural of Law and Economics*, **5**, 1.

Struck, Myron (1982), 'Ring in the New Year and wring out the Old', *Washington Post*, October.

Sunstein, Cass R. (1987), 'Constitutionalism after the New Deal', *Harvard Law Review*, **101**, 421.

Tarlock, A. Dan (1994), 'The non-equilibrium paradigm in ecology and the partial unraveling of environmental law', *Loyola of Los Angeles Law Review*, **27**, 1121.

Tarlock, A. Dan (2000), 'The future of environmental "rule of law" litigation', *Pace Environmental Law Review*, **17**, 237.

Teclaff, Ludwik A. (1967), *The River Basin in History and Law*, The Hague: Martinus Nijhoff.

Teclaff, Ludwik A. (1991), 'Treaty practice related to transboundary flooding', *Natural Resources Journal*, **31**, 109.

Trudgill, Stephen T. (1988), *Soil and Vegetation Systems*, 2nd edn, New York: Oxford University Press.

Turner, James S. (1970), *The Chemical Feast*, New York: Penguin Books.

Volkman, John M. (1999), 'How do you learn from a river? Managing uncertainty in species conservation policy', *Washington Law Review*, **74**, 719.

Wengert, Norman (1981), 'A critical review of the river basin as a focus for resources planning, development, and management', in Ronald M. North et al. (eds), *Unified River Basin Management 9*, Minneapolis: American Water Resources Association

Westbrook, David A. (1994), 'Liberal environmental jurisprudence', *University of California Davis Law Review*, **27**, 619.

Winter, Ralph K. (1973), 'Economic regulation and competition: Ralph Nader and creeping capitalism', *Yale Law Journal*, **82**, 890.

Worster, Donald (2001), *A River Running West*, New York: Oxford University Press.

Wright, Patrick (2001), 'Fixing the delta: the CALFED Bay-Delta Program and water policy under the Davis Administration', *Golden Gate University Law Review*, **31**, 331.

Western Water Policy Review Advisory Commission (WWPRAC) (1988), *Water in the West: Challenge For the Next Century*, Denver: US Department of the Interior.

Xi, Wang, Zhang Xiaobo, Li Wenkai, Gu Dejin and Zhou Yanfang (2002), 'Managing water resources for a sustainable future: law, policy, and methodology of China', paper prepared for the Natural Resources Law Center, University of Colorado School of Law, Boulder, CO.

Index

aboriginal water rights (Australia) 122–4
acequias 41–2
adaptive management 13, 47, 63, 88
Ahmedabad, India 57
Ak-Chin Indian Community 111–12
allotment policy 102, 106
American Convention on Human Rights 125
Anderson–Cottonwood Irrigation District 81
Angola 146
Animas–La Plata Project 112
Annan, Kofi (UN Secretary General) 142
Aral Sea 144
area of origin protections 75–6, 96
Argentina 51, 54, 61
Australia 4, 6, 30, 48–51, 61, 85–9, 94, 96–9, 122–4, 126, 157

Babbitt, Bruce 63, 118
BATNA 149
beneficial use 51–2, 77, 81, 172, 174
benefit-cost analysis 32–3
Big Horn River (and Native American water rights) 110, 119
Black Sea 14
Bolivia 11, 41
Bonneville Power Administration 134
Botswana 146
Boulder Canyon Project Act 165
Brazil 30, 38, 61
British Columbia Power and Hydro Authority 134
Buenos Aires Water Concession 54, 61
Bush, George W. (administration of) 170–71

CALFED 13, 37–8, 47, 63, 84, 173
Cambodia 145

Canada 4, 6, 30, 48, 133–5, 141, 151
Central Arizona Project 112
Central Valley Project Improvement Act 38, 83, 96
Chennai, Thailand 57–8
Chile 11, 26, 52–3, 61
China 4, 7, 16, 85–6, 91–4, 97–8, 143–5, 157
Clean Water Act 29, 79–80, 94, 163, 168, 169–72
climate change 14–15, 95, 159
Clinton, William (administration of) 118, 169
Cochabamba, Bolivia (and water privatization) 41, 97, 99
Colorado River Basin Salinity Control Act 136
Colorado River 30–31, 38, 45–6, 96, 107–8, 112, 114, 119, 135–41
Columbia River 46, 134, 141, 151, 173
Commerce Clause 76
Commission on Environmental Cooperation 138
Commission on Sustainable Development 161–2
compact commissions 139–40
comparative water law 3, 10–11, 85
compensatory storage 76, 96
comprehensive coordinated joint plans (CCJPs) 166
conjunctive use and management 37, 63
Convention on Wetlands (Ramsar Convention) 86
cultural value of water 29, 102

dam removal 82
Damodar Valley Authority 5
Danube River 14, 31
Declaration of Principles of Indigenous Rights 127
Denver Water Board 44

diffusion of water policy innovations 3,
    12–13

East–West Center 6
ecological integrity paradigm 157
ecologicalism 159, 162–3
economic efficiency (in water use) 32–3
Edwards Aquifer 32, 37
Egypt 4–5
Endangered Species Act 29, 47, 80–81,
    84, 112, 150, 163
England 5–6, 9, 11
English common law 5, 77
environmental determinism 9
Environmental Impact Statement 78
Environmental Protection Agency 36,
    98, 166, 171–2
environmental regulation 78–83, 94,
    166–8, 173
equitable apportionment 138
equity (in water allocation and realloca-
    tion) 33–34, 51, 60, 89–90, 95–6,
    99
European Union 9

Federal Energy Regulatory Commission
    81–2, 98
Federal Power Act 81–2, 165
federal reserved water rights 46, 78,
    104–110, 114, 118, 120
Federal Water Pollution Control Act
    168–9
federalism 73, 78–83, 98, 103, 109,
    116–17
Fish and Wildlife Coordination Act 46,
    81
Flood Control Act 45–6
Fort Belknap Indian Reservation 106
France 4, 6, 9, 11

Gal Oya Authority 5
Ganges River 31
General Agreement on Tariffs and Trade
    (GATT) 137
general stream adjudications 110–11,
    115
Germany 4, 9, 51
Glen Canyon Dam 82
Global Alliance for Water Security 151
Global Environmental Facility 14

Global Water Partnership 7, 10
Grand Canyon Protection Act 82
Great Britain 4
Great Lakes 133–4
Green Revolution 12

Habitat Conservation Plans (HCPs) 163
Hagerstrand, Torsten 13
Harmon Doctrine 136, 150
Hawaii Commission on Water Resources
    Management 26
Hoover Dam 46

Imperial Irrigation District 28, 40
India 4–5, 8, 16, 57, 143–4, 151, 157
Indian water rights
    change of use 118–20
    origins, *see* federal reserved water
        rights
    quantification of 108–110, 114, 118,
        120
    settlements 111–13
indigenous water rights protections in
    international law 125–7
Indus Waters Treaty 150–51
Inland Waterways Commission 164
instream flows 29, 77, 89
Integrated Water Resources Management
    (IWRM) 155–6, 159–63, 172, 174
Integrated Watershed Management
    (IWM) 156, 159, 163–4, 170–74
International Boundary and Waters
    Commission 135–7
international comparative studies 3–6
International Convention on the
    Elimination of All Forms of Racial
    Discrimination 125
International Court of Justice 127, 149
International Covenant on Civil and
    Political Rights 125–6
International Covenant on Economic,
    Social and Cultural Rights 125
International Geosphere Biosphere
    Program 159
International Institute of Applied
    Systems Analysis 6
International Joint Commission 133, 141
International Labour Organization
    Convention No. 169, 125
International Rivers Network 15

International Water Law Association 10
International Water Management
    Institute 8
International Water Resources Associa-
    tion 7, 10
Internet water resources information 8
interstate water compacts 30, 45,
    138–140
interstate water marketing 30–31, 38,
    48–51
intertemporal externalities 31–2
Iran 6, 143
Iraq 6
Islamic water law 11
Israel 6, 147
Italy 4–7, 9

James, William 11
Japan 7, 86, 126
Jordan River 150
jurisdictional externalities 30, 38, 170

Katmandu, Nepal 55–7, 59
Kinney, Clesson S. 4
Klamath River 150
Klamath Tribe 118
Kousou Dam 34

Lake Constance 54–5
Lake Mead 31
Laos 145, 151
League of Nations 5
legal transplant theory 16–17
Lindblom, Charles 162
Los Angeles Water and Power 44
Lowi, Theodore 168

Marsh, George Perkins 4
McCarren Amendment 30, 109–11,
    114–15
Mekong River Commission 145, 149
Mekong River 145, 151
Melamchi Project 56
Mescalero Apache Tribe 117
Metropolitan Water District of Southern
    California 28, 37, 40
Mexico 4, 6, 31, 51, 135–7
Ministerial Declaration of the Interna-
    tional Conference on Freshwater
    162

Mono Lake 76–7, 90, 124
Montesquieu, Baron de 8, 16
multiscalar studies 140–41
Murray–Darling Basin Commission 26,
    48–51, 86, 88–9, 97, 99
Myanmar (Burma) 143–5

Namibia 146
National Environmental Policy Act 46,
    78–9, 166
National Marine Fisheries Service 80,
    98
National Planning Board 165
National Pollution Discharge Elimina-
    tion System (NPDES) 169
National Research Council 6
National Resources Committee 165
National Resources Planning Board 165
National Water Act (of South Africa)
    89–91, 125
National Water Commission 107
National Watershed Commission 172
National Waterways Commission 164
Native Title Act (of Australia) 122
Nepal Water Supply Corporation 56–7
Nepal 55–7, 59, 151
New South Wales Water Management
    Act 88
Newlands Commission 164
Nile River 6, 31
no injury rule 29, 36, 72, 119–20
non-equilibrium paradigm 158–60
North American Free Trade Agreement
    (NAFTA) 137–8
Northern Colorado Water Conservancy
    District 27–8, 40, 48
Northwest Power Planning Council 140

Ogallala Aquifer 31, 35–6
Okavango River 146
Ottoman Empire 4

Pakistan 143–4
Palestinian Authority 147
Palo Verde Irrigation District 28, 40
Papago Indian Reservation 112
Pecos River 40
Permanent Okavango River Basin
    Commission 146
Peru 5

population (and water supplies) 142
Powell, John Wesley 164
practicably irrigable acreage (PIA)
        108–109, 114
precautionary principle (in Australian
        water law) 87, 94
President's Water Resources Policy
        Commission 5
prior appropriation 28, 43, 71–2, 86, 93,
        95, 108, 174
private and cooperative water develop-
        ment 43–4
private sector (defined) 26
privatization (of water) 11, 26, 33, 39,
        40–42, 52–4, 60
Progressive Era 164–5
public goods 41
public interest protection and review
        73–5, 120, 124–5, 172
public interests and values of water
        (defined) 69, 72–3
public sector (defined) 25–6
public trust doctrine 76–7, 89, 124–5,
        172
Pyramid Lake Paiute Tribe 107, 114–15

Racial Discrimination Act (of Australia)
        122
Reagan, Ronald (administration of) 168
reallocation of water 27–8, 88, 96, 99,
        173
Reclamation Act 150, 165
recreational water rights 29–30
Resources Conservation and Recovery
        Act 94
Rhine River 31, 54–5
Rio Grande/Rio Bravo River 31, 135,
        138, 173
riparian water law 70–71, 85–6, 90,
        174
river basin development and administra-
        tion 45, 60–61, 85, 91–2, 94,
        134–7, 141, 151, 155–6, 159–74
Rogers, Everett 13
Roman water law 4–5, 9, 16
Roosevelt, Franklin (administration of)
        165
Roosevelt, Theodore (administration of)
        46, 164
rule of capture 27, 71

rule of law ideology 166

salinity (along the US–Mexico border)
        136
Salween River 143–4
Sardar Sarovar Project 34
Scotland 6–7
security studies 131–2
Semple, E.C. 16
Snake River 28, 173
social learning 3, 13–15
social movement theory 3, 14–15
South Africa 6, 15, 85, 89–91, 94, 96,
        98–9, 124–5, 157
South Platte River 32, 37, 139
Soviet Union 6, 144
Spain 1, 6, 9, 42, 51
Spanish missions 4
Stockholm Declaration 126
subsidies 27, 58, 62, 71, 88, 96
Superfund (CERCLA) 94
sustainable development 35, 87, 155,
        157–60, 163–4, 172–3

takings 88, 96
Tarde, Gabriel de 12
Tennessee Valley Authority 5
Thailand 143–5, 151
third parties 73–4, 84, 97
total maximum daily load (TMDL) 37,
        79, 169–72
transactions costs 51
Transboundary Freshwater Disputes
        Database 8, 143–4
Treaty of Guadalupe Hidalgo 135
Treaty Relating to Boundary Waters
        between the United States and
        Canada 133
tribal water codes 116–18
Trinational Permanent Water Commis-
        sion 146
Truckee River 107
Tucson Water 44

Unified River Basin Management
        (URBM) 156, 164–70
United Kingdom 6, 133
United Nations Conference on Environ-
        ment and Development 126, 156,
        158, 160–61

United Nations Convention for the
Protection of the World Culture and
Natural Heritage 86
United Nations Convention on Biologi-
cal Diversity 86, 125–26
United Nations Draft Declaration on
the Rights of Indigenous Peoples
125
United Nations General Assembly 162
United Nations 7
United States Army Corps of Engineers
5, 44–6, 80, 85, 98, 134, 165
United States Bureau of Reclamation 5,
7, 28, 31, 36, 44–5, 82, 85, 106–
107, 112, 139, 165
United States Fish and Wildlife Service
80, 98
United States Geological Survey 5
United States National Intelligence
Council 142
United States Soil Conservation Service
5
United States-Mexico Water Treaty
135–6
usufructuary rights 71

Vienna Declaration 126
Vietnam 145

Ware, Eugene 4
water banks 26, 40, 48, 62
water entitlements 172–74
water markets and marketing, transfers
26, 33, 38–42, 48–51, 56–9, 62,
72–6, 89, 96–7, 99, 113, 115–16
water pricing 28–9, 89–92, 97, 99

water quality 36–7, 55–8, 63, 79, 92,
133; *also see* Clean Water Act
Water Resources Council 166, 168–9
Water Resources Planning Act 166, 169
water rights administration 71
water rights certainty 95, 172–74
water scarcity 27, 70, 142–3, 157
water stress 143
water trusts 77
watershed groups 26, 47, 61, 83–4
Watson, Alan 16–17
Weber, Kenneth 42
West Bank 147
Western Water Policy Review Advisory
Commission 6, 160
wetlands protection 80
Wild and Scenic Rivers Act 82
Wind River Tribes 110, 119
Winters doctrine, *see* Indian water
rights, federal reserved water rights
Wittfogel, Karl 9
World Bank 7, 41, 53
World Climate Research Program 159
World Commission on Dams 6, 15
World Commission on Environment and
Development (Brundtland Commis-
sion) 35
World Council of Indigenous Peoples
127
World Heritage Convention 126
World Summit on Sustainable Develop-
ment (WSSD) 156, 158, 160–61
World Water Commission 7
World Water Council 10

xeriscaping 28